SpringerBriefs in Statistics

SpringerBriefs present concise summaries of cutting-edge research and practical applications across a wide spectrum of fields. Featuring compact volumes of 50 to 125 pages, the series covers a range of content from professional to academic. Typical topics might include:

- A timely report of state-of-the art analytical techniques-
- A bridge between new research results, as published in journal articles, and a contextual literature review
- A snapshot of a hot or emerging topic
- An in-depth case study or clinical example
- A presentation of core concepts that students must understand in order to make independent contributions

SpringerBriefs in Statistics showcase emerging theory, empirical research, and practical application in Statistics from a global author community.

SpringerBriefs are characterized by fast, global electronic dissemination, standard publishing contracts, standardized manuscript preparation and formatting guidelines, and expedited production schedules.

Monia Lupparelli · Giovanni Maria Marchetti ·
Claudia Tarantola

Regression Graph Models for Categorical Data

Parameterization and Inference

 Springer

Monia Lupparelli
Department of Statistics, Computer
Science, Applications
University of Florence
Florence, Italy

Giovanni Maria Marchetti
Department of Statistics, Computer
Science, Applications
University of Florence
Florence, Italy

Claudia Tarantola
Department of Economics, Management
and Quantitative Methods
University of Milan
Milan, Italy

ISSN 2191-544X   ISSN 2191-5458 (electronic)
SpringerBriefs in Statistics
ISBN 978-3-031-99796-9   ISBN 978-3-031-99797-6 (eBook)
https://doi.org/10.1007/978-3-031-99797-6

Mathematics Subject Classification: 62H17, 62F15

This Springer imprint is published by the registered company Springer Nature Switzerland AG
The registered company address is: Gewerbestrasse 11, 6330 Cham, Switzerland

If disposing of this product, please recycle the paper.

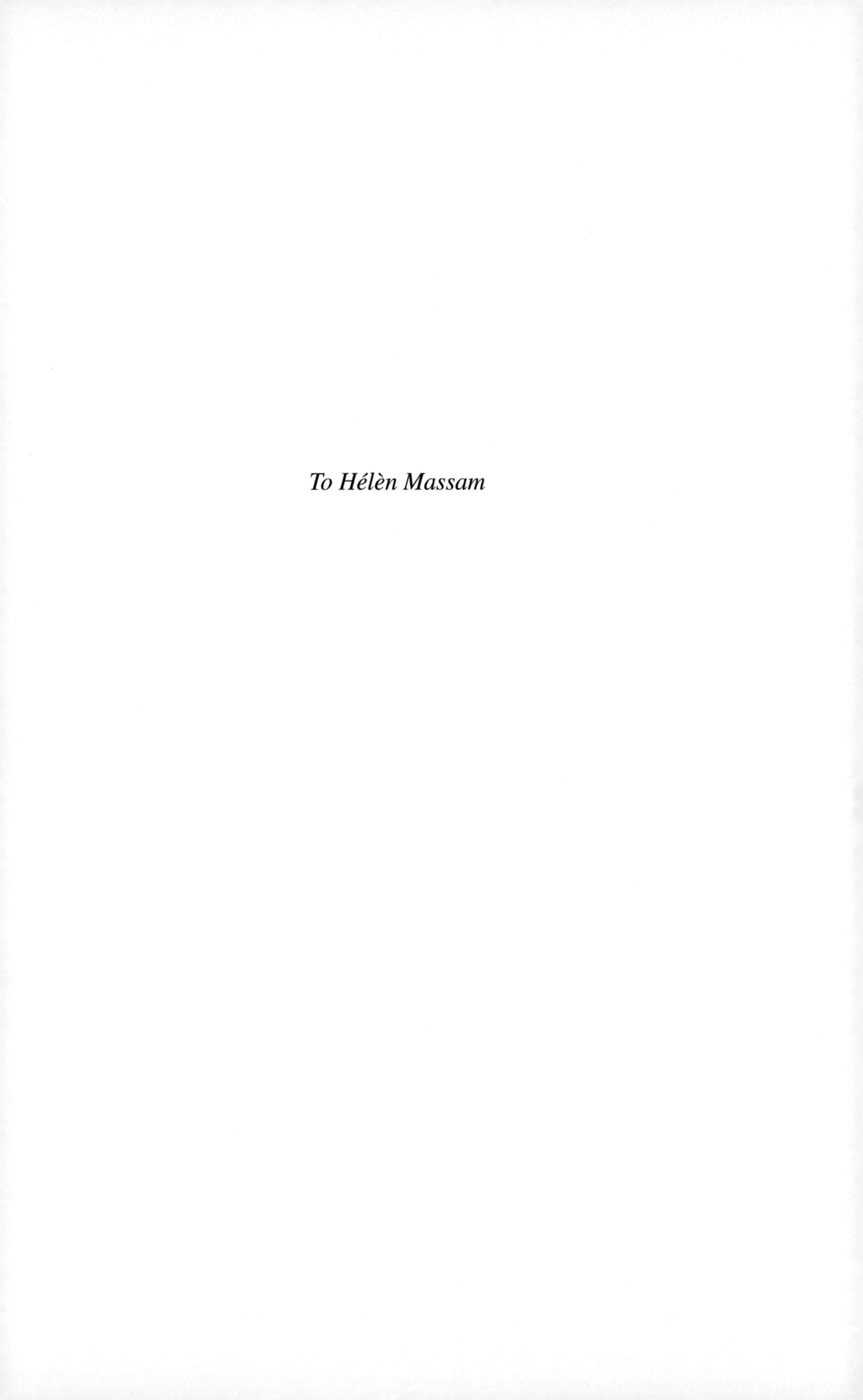

To Hélèn Massam

Preface

In this work we have attempted to summarize our activity in the field of chain graph models, with a particular emphasis on the parameterizations and inference of regression chain graph models, often referred as regression graph models, for the analysis of categorical data. Unlike other families of graphical models, such as the Bayesian networks (that is directed acyclic graphs) or the undirected graph models, which are by now extensively known and used, the regression graphs do not enjoy yet the same popularity. Our motivation for writing this book stems from the conviction that this class of models can be highly beneficial for statistical data analysis. This family of graphs defines a framework of recursive multivariate regressions that find application in important fields such as structural equation models and marginal models for qualitative data. In contrast to undirected graphical models, the regression graph models allow the variables considered on equal standing to be grouped into chain components that can serve different roles in the data-generating mechanism. Regression graphs also bridge some limitations of Bayesian networks used to define univariate recursive regression models that do not account for additional forms of dependencies for instance induced by unobserved variables.

This work has a twofold intention. The first is to consolidate knowledge on regression graphs, their interpretation in terms of sequences of multivariate regressions, interpretable parameterizations for categorical data, and inference and model selection within the frequentist and Bayesian approaches. The second objective of this work is encouraging both the applications and the research concerning this class of graphical models.

The book is intended primarily for graduate and Ph.D. students in Statistics and Data Science who are familiar with the basics of graphical Markov models and of categorical data analysis, and for motivated researchers in specific applied fields. Data and R code used in the book are made available at the website https://github. com/StaThin/RGM.

The writing of this book has been stimulated and enriched by collaborations and discussions over the time with colleagues and friends from the community of graphical models and categorical data analysis. We are especially grateful to Nanny Wermuth, David Cox, Wicher Bergsma, Tamás Rudas, Ioannis Ntzoufras,

Anna Gottard, Kayvan Sadeghi, Alberto Roverato, Luca La Rocca, Mathias Drton, Robin Evans, Francesco Bartolucci, Antonio Forcina, Guido Consonni, Søren Højsgaard. We would also thank the Editors of Springer Publications and the Anonymous Reviewers for their careful reading, constructive comments and suggestions that have thoroughly improved this manuscript.

Florence, Italy   Monia Lupparelli
June 2025   Giovanni Maria Marchetti
                                                                                      Claudia Tarantola

Competing Interests The authors have no competing interests to declare that are relevant to the content of this manuscript.

Contents

Chapter 1
Regression Graph Models

Summary

This chapter introduces a class of graphical Markov models broadly called *regression graph models*. In general all graphical models are defined by specific conditional independence constraints in a system of random variables, constraints that have a precise graph representation. These models are useful to specify stepwise data generating processes and research hypotheses in cohort studies. We give some definitions and properties common to all graphical models, and then we discuss some examples. We will explain in some detail the case of a system of Gaussian variables.

1.1 Graphical Markov Models

1.1.1 Introduction

The main idea behind the statistical models discussed in this book is the fact that any collection of random variables $X = (X_1, \ldots, X_d)$ can be associated to the vertices of some graph with properties that can be used to define precise constraints on the joint distribution of the variables. A graph in general is a mathematical object defined by a set of vertices $V = \{1, \ldots, d\}$ and a set of edges $\mathcal{E}$ connecting pairs of vertices. There are three main classes of graphs (with no self loops nor multiple edges) having the same type of edges: the *undirected graphs* with full line edges, the *directed acyclic graphs* with directed edges but no cycles and the *bi-directed graphs* with bi-directed edges. The three classes are shown in Fig. 1.1.

We start from some basic facts concerning the concept of *independence* of random variables that are essential for graphical models. We will discuss only sets of discrete or absolutely continuous variables. If X is a discrete variable with levels $x \in \mathcal{X}$ we denote by $p(x) = P(X = x)$ its probability mass function and we use the same

© The Author(s), under exclusive license to Springer Nature Switzerland AG 2025

M. Lupparelli et al., *Regression Graph Models for Categorical Data*,

SpringerBriefs in Statistics, https://doi.org/10.1007/978-3-031-99797-6_1

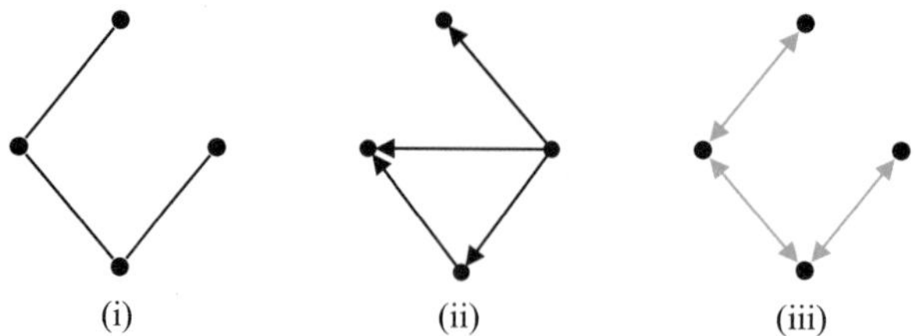

(i)   (ii)   (iii)

Fig. 1.1 **i** An undirected graph; **ii** A directed acyclic graph; **iii** A bi-directed graph

notation for the probability density function (even if it is dimensionally different) when X is a continuous variable. This simplified notation is extended to several discrete or several continuous variables.

Assume that we have three discrete random variables X_1, X_2, X_3 with a joint distribution denoted by $p_{123}(x_1, x_2, x_3)$. Often we omit the arguments of the function using the short form p_{123}, and we will assume that $p_{123} > 0$. From the joint distribution we obtain all the marginal distributions (e.g., p_1, p_{23} and so on) and the conditional distributions (e.g., $p_{1|23} = p_{123}/p_{23}$).

Recall that the joint distribution always admits a *factorization* according to a specific order of the variables. For example,

$$p_{123} = p_{1|23}\,p_{2|3}\,p_3. \tag{1.1}$$

There are two quite distinct concepts of independence between variables: the marginal and the conditional independence.

1.1.2 Marginal Independence

The variables X_1 and X_2 are said to be *marginally independent*, written $X_1 \perp\!\!\!\perp X_2$, if

$$p_{12}(x_1, x_2) = p_1(x_1)p_2(x_2) \text{ for any } x_1, x_2, \text{ short form } p_{12} = p_1 p_2. \tag{1.2}$$

Equivalently, $X_1 \perp\!\!\!\perp X_2$ if and only if

$$p_{1|2}(x_1|x_2) = p_1(x_1) \text{ for any } x_1, x_2, \text{ short form } p_{1|2} = p_1. \tag{1.3}$$

A marginal independence can be interpreted by saying that X_2 is irrelevant to predict X_1 because the conditional distribution $p_{1|2}$ is actually not a function of x_2. Notice that it extends to transformations of the variables because $X_1 \perp\!\!\!\perp X_2$ implies that $g(X_1) \perp\!\!\!\perp h(X_2)$ for any functions $g(\cdot)$ and $h(\cdot)$.

A marginal independence implies a simplification of the factorization (1.1). For instance, under the independence $X_1 \perp\!\!\!\perp X_2$ the joint probability of the three variables results in

$$p_{123} = p_{3|12} p_1 p_2.$$

The definitions of independence are extended from the discrete to the continuous variables substituting the probability functions with the density functions.

1.1.3 Conditional Independence

The concept of conditional independence involves instead at least three variables. We say that the variables X_1 and X_2 are *conditionally independent given* X_3 if, for any x_3, $X_1 \mid (X_3 = x_3)$ and $X_2 \mid (X_3 = x_3)$ are independent. Using the short form, this means that

$$p_{12|3} = p_{1|3} p_{2|3} \tag{1.4}$$

and is denoted by $X_1 \perp\!\!\!\perp X_2 \mid X_3$. It is not difficult to show that an equivalent condition is

$$p_{1|23} = p_{1|3},$$

that is the conditional distribution of X_1 given $X_2 = x_2$ and $X_3 = x_3$ does not depend on the value of x_2.

The conditional independence $X_1 \perp\!\!\!\perp X_2 \mid X_3$ implies that the joint density can be factorized in 3 different ways as

$$p_{123} = p_{1|3} p_{2|3} p_3 = p_{2|3} p_{3|1} p_1 = p_{1|3} p_{3|2} p_2.$$

This shows that the concepts of marginal and conditional independence are quite different and for instance we can have

$$X_1 \perp\!\!\!\perp X_2 \text{ but } X_1 \not\perp\!\!\!\perp X_2 \mid X_3 \text{ and } X_1 \perp\!\!\!\perp X_2 \mid X_3 \text{ but } X_1 \not\perp\!\!\!\perp X_2.$$

1.1.4 Role of the Variables

The above purely probabilistic facts have an important function in the interpretation of variables in statistical studies. Typically an investigation is based on n observations collected on many variables $X_1, X_2, \ldots, X_d$ that can be distinguished as

- purely explanatory, (background) variables
- intermediate variables
- response variables.

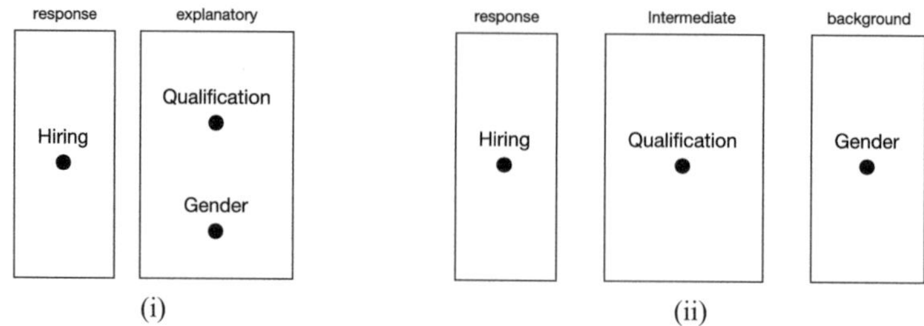

Fig. 1.2 Two different orderings of the variables. **i** Two blocks: one response and two explanatory variables. **ii** Three blocks: one response, one intermediate variable and one background variable

Example 1.1 (Labor market) Consider a study on labor market where 3 binary variables are observed:

- Hiring: the successful job placement of an individual (Yes = 1, No = 0).
- Qualification: the university studies of the applicant (Classical, Scientific).
- Gender of the applicant (F = female, M = male).

If we want to predict the binary response Hiring from Qualification and Gender we have two blocks of variables: the first contains the response and the second the explanatory variables as in Fig. 1.2. However, we can also consider a different ordering of the variables in three blocks determined by subject-matter considerations. Namely we can consider Hiring as the primary response depending on both the Qualification and Gender, Gender is a purely background variable and Qualification is an intermediate variable because it influences Hiring but is itself a response for Gender.   □

1.2 Marginal Independence Models

Graphical Markov models are statistical models that define specific conditional independence constraints for a set of random variables $X_1, X_2, \ldots, X_d$, with a precise graphical representation. There are several types of graphical models, each defining a specific interpretation not only in terms of independence but also of dependencies among the variables.

It is important to distinguish two different meanings of dependence: the undirected and the directed dependence. The first is the association between two variables assumed on an equal standing and the second is the relation between a response and an explanatory variable. The above distinction requires some knowledge of the system under study and is often provisional, but it is essential in defining the possible orders of the variables.

Moreover, another distinction will be essential: that between marginal and conditional independence. A marginal independence between two variables is observed within their marginal distribution ignoring all the other variables. A conditional independence is instead found for the two variables conditional on a given set of other variables.

The main idea of graphical models is that of representing the variables $X_1, \ldots, X_d$ as dots, and the dependencies among them with links between the dots. This kind of representation is called a *graph* where the dots are called *vertices* or *nodes* and the links are called *edges*. The graph is denoted by $\mathcal{G} = (V, \mathcal{E})$ where V is the set of vertices and $\mathcal{E}$ is the set of vertices.

If there is some kind of independence between two variables then this is represented by a missing edge between the two dots. This representation is called an *independence graph* because the missing edges indicate an independence.

In this chapter we will introduce first a type of independence graph sometimes called *covariance graph*. We will consider first some examples with a few random variables having a joint multivariate Gaussian distribution, while in Chap. 2 we will give an in-depth treatment of the case of binary or categorical variables.

Example 1.2 (Covariance graphs) Let $X = (X_1, X_2, X_3)^T$ be a random vector with a Gaussian distribution with mean vector and covariance matrix, respectively,

$$\boldsymbol{\mu} = \begin{bmatrix} \mu_1 \\ \mu_2 \\ \mu_3 \end{bmatrix}, \quad \boldsymbol{\Sigma} = \begin{bmatrix} \sigma_{11} & \sigma_{12} & \sigma_{13} \\ \cdot & \sigma_{22} & \sigma_{23} \\ \cdot & \cdot & \sigma_{33} \end{bmatrix}. \tag{1.5}$$

The joint density function p_{123} describes the behavior of the three variables considered on an equal standing.

Now consider the model of marginal independence between X_2 and X_3, i.e., $p_{23} = p_2 p_3$. It is well-known that for a Gaussian distribution this is equivalent to the constraint $\sigma_{23} = 0$. This model is the family of Gaussian distributions with a positive definite covariance matrix containing two zeros in positions $(2, 3)$ and $(3, 2)$. To represent this model we choose a bi-directed graph with three vertices $V = \{1, 2, 3\}$, two bi-directed edges $1 \leftrightarrow 2$ and $1 \leftrightarrow 3$ and a missing edge between the vertices 2 and 3 as in Fig. 1.3i. The bi-directed edges are also called simply *arcs*. The joint probability distribution is denoted also by p_V where $V = \{1, 2, 3\}$ are the vertices of the bi-directed graph. $\qquad \square$

The graph therefore describes the model of marginal independence between the variables X_2 and X_3 with a missing arc between 2 and 3, by implicitly showing the

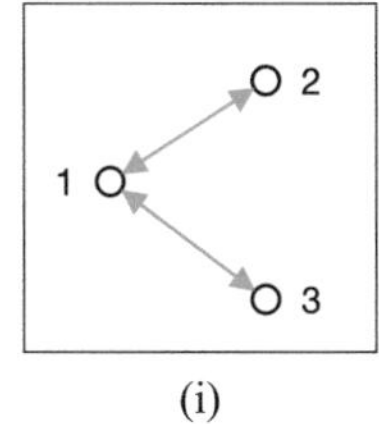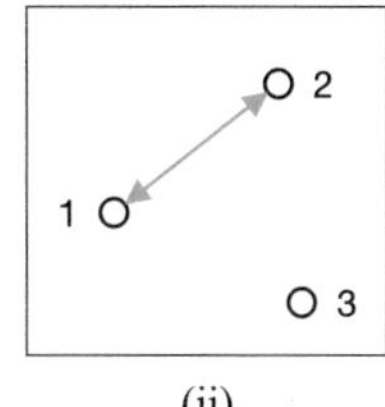

Fig. 1.3 Two covariance graphs. **i** Marginal independence $X_2 \perp\!\!\!\perp X_3$. **ii** Joint independence $(X_1, X_2) \perp\!\!\!\perp X_3$

dependence structure with the arcs present. The justification of the use of bi-directed edges will be clearer later on.

Such *bi-directed graphs* are sometimes called also *covariance graphs* because the independence constraints concern only the covariance parameters. Note that all the edges of the covariance graphs are by definition bi-directed and that the labels of the nodes are just the indices of the variables.

Consider now the covariance graph in Fig. 1.3b with two missing edges $1 \leftrightarrow 2$ and $2 \leftrightarrow 3$.

The two missing edges are interpreted again as two independencies $X_1 \perp\!\!\!\perp X_3$ and $X_2 \perp\!\!\!\perp X_3$ using a rule called the *pairwise Markov property*. For a jointly Gaussian distribution this implies the vanishing of two covariances $\sigma_{13} = \sigma_{23} = 0$, but it implies also a stronger *joint independence* $(X_1, X_2) \perp\!\!\!\perp X_3$, meaning that

$$p_{123} = p_{12} p_3.$$

Example 1.3 (Lawyer's ratings of state judges) Data compatible with the above covariance graph are `USJudgeRatings` from the R package **datasets** containing Lawyers' ratings of state judges in the US Superior Court. The variables X_1 (judicial integrity), X_2 (diligence) and X_3 (number of contacts of lawyer with judge) were observed on 43 judges. The data appear compatible with a Gaussian distribution with sample correlation matrix

$$\begin{bmatrix} 1 & 0.872 & -0.133 \\ . & 1 & 0.012 \\ . & . & 1 \end{bmatrix}$$

and it can be verified that a joint test of $\rho_{13} = 0$ and $\rho_{23} = 0$ is not significant, suggesting the independencies $X_1 \perp\!\!\!\perp X_3$ and $X_2 \perp\!\!\!\perp X_3$. $\qquad\square$

1.2.1 The Connected Set Markov Property

Assuming that the variables are jointly Gaussian, like in the previous example, we expect that the implication

$$(X_1 \perp\!\!\!\perp X_3) \text{ and } (X_2 \perp\!\!\!\perp X_3) \implies (X_1, X_2) \perp\!\!\!\perp X_3, \tag{1.6}$$

called *composition property*, holds. But it should be stressed that this property is not valid in general for all distributions, as illustrated by the following example.

Example 1.4 (A distribution violating the composition property) Consider the trivariate binary distribution of Table 1.1. You can verify that the two marginal tables (X_1, X_3) and (X_2, X_3) are uniform, implying the two independencies $X_1 \perp\!\!\!\perp X_3$ and $X_2 \perp\!\!\!\perp X_3$. However surprisingly the joint independence $(X_1, X_2) \perp\!\!\!\perp X_3$ does *not* hold. This appears rather strange, and it would seem odd to represent the marginal independence structure of this distribution with the graph of Fig. 1.3ii because the

Table 1.1 A probability table that does not satisfy the composition property

	X_3	0		1		
X_1	X_2	0	1	0	1	Sum
0		0.20	0.05	0.05	0.20	0.5
1		0.15	0.10	0.10	0.15	0.5
Sum		0.35	0.15	0.15	0.35	1.0

vertex 3 is totally disconnected from the set of vertices $\{1, 2\}$ while the corresponding variable X_3 is not jointly independent of (X_2, X_3). $\qquad\square$

The example suggests that it is convenient to have a rule to read the independencies off the graph that, unlike the pairwise Markov property, prevents the violation of the composition property. One such rule is called the *connected sets Markov property*; see Richardson (2003).

This property requires some graph concepts explained below. We assume that $\mathcal{G}(V, E)$ is a bi-directed graph and that A is a nonempty subset of the vertices V.

- The *subgraph* induced by A is the graph denoted by $\mathcal{G}_A$ with vertex set A and with edges that have both endpoints in A. For example, given the graph of Fig. 1.3ii, the subgraph induced by $A = \{1, 2\}$ has vertex set $\{1, 2\}$ and a single edge $1 \leftrightarrow 2$.
- A *path* in a bi-directed graph is a sequence of edges which joins a sequence of vertices which are all distinct. For example, in the graph of Fig. 1.3i, $1 \leftrightarrow 2 \leftrightarrow 3$ is a path connecting the vertices 1, 2 and 3.
- The subgraph induced by A is said *connected* if every pair of vertices in A is joined by a path on which every vertex is in A. Otherwise, it is said *disconnected*.
- The *set of all the disconnected subgraphs* of $\mathcal{G}$ is the class $\mathcal{D}$ composed by all the subgraphs induced by the subsets of the vertex set that are disconnected.
- Every disconnected subgraph of $\mathcal{G}$ may be expressed uniquely (up to order) as a disjoint union of connected graphs called its *connected components*.

The class of all the disconnected subgraphs of the graph of Fig. 1.3ii and their connected components is shown in Table 1.2. Then the connected set Markov property can be expressed as follows; see Lupparelli et al. (2009).

Definition 1.1 (*Connected set Markov property*) Let $\mathcal{G} = (V, \mathcal{E})$ be a bi-directed graph. Then, the joint probability distribution p_V of a random vector X_V satisfies the *connected set Markov property* with respect to $\mathcal{G}$ if and only if for every disconnected subgraph of $\mathcal{G}$ with connected components with vertex sets $\{A_1, \ldots, A_r\}$ it turns out that $X_{A_1} \perp\!\!\!\perp \ldots \perp\!\!\!\perp X_{A_r}$.

Table 1.2 All the disconnected subgraphs of the graph of Fig. 1.3ii and the corresponding connected components

Disconnected subgraphs	$\mathcal{G}_{2,3}$	$\mathcal{G}_{\{1,3\}}$	$\mathcal{G}_{\{1,2,3\}}$
Connected components	$\{2\}, \{3\}$	$\{1\}, \{3\}$	$\{1, 2\}, \{3\}$

The interpretation of the graph of Fig. 1.3i according to the connected set Markov property is that the unique disconnected subgraph is $\mathcal{G}_{2,3}$ with connected components $\{2\}$ and $\{3\}$ implying the marginal independence $X_2 \perp\!\!\!\perp X_3$. On the other hand the interpretation of the graph of Fig. 1.3ii follows by listing all its disconnected subgraphs and their components, as in Table 1.2, and translating them into the marginal independencies $X_1 \perp\!\!\!\perp X_3$, $X_2 \perp\!\!\!\perp X_3$ plus the joint independence $(X_1, X_2) \perp\!\!\!\perp X_3$.

A consequence is that the distribution of Example 1.4 cannot be represented by the graph of Fig. 1.3ii under the connected set Markov property, because the property automatically implies the composition property. In general, it is important to acknowledge that in the family of all the cross-classified probability tables the composition property can be violated. Nevertheless, it is possible to define graphical models that satisfy the composition property using the connected set Markov property.

1.2.2 Conditional Covariance Graphs

We have seen that the covariance graphs are models of marginal independence for a set of variables considered on an equal standing. We consider now covariance graphs for a set of variables *conditionally* on a set of explanatory variables. Again we will discuss first the case of variables with a multivariate Gaussian distribution.

Example 1.5 (Conditional covariance graph models) Given three normal variables X_1, X_2, X_3 defined as in Eq. (1.5), we are interested in the conditional distribution $p_{12|3}$ of X_1, X_2 given X_3. As multivariate Gaussian distributions are closed with respect to conditioning and marginalization the density $p_{12|3}$ is a bivariate normal distribution with parameters

$$\boldsymbol{\mu}_{a|b} = \begin{bmatrix} E(X_1 \mid X_3 = x_3) \\ E(X_2 \mid X_3 = x_3) \end{bmatrix} \qquad \Sigma_{aa|b} = \begin{bmatrix} \sigma_{11|3} & \sigma_{12|3} \\ \cdot & \sigma_{22|3} \end{bmatrix} \tag{1.7}$$

where $a = \{1, 2\}$, $b = 3$, $\boldsymbol{\mu}_{a|b}$ is the vector of conditional expectations and $\Sigma_{aa|b}$ is the conditional covariance matrix. The sets a and b denote respectively the set of the two responses and the set of the single explanatory variable.

The conditional expectations, are the linear *regression functions* that can be expressed in terms of the original parameters as follows

$$\begin{aligned} E(X_1 \mid X_3 = x_3) &= \mu_1 + \beta_{13}(x_3 - \mu_3) \\ E(X_2 \mid X_3 = x_3) &= \mu_2 + \beta_{23}(x_3 - \mu_3) \end{aligned} \tag{1.8}$$

where $\beta_{13} = \sigma_{13}/\sigma_{33}$ and $\beta_{23} = \sigma_{23}/\sigma_{33}$ are called marginal *regression coefficients*. The conditional covariance can be written as

$$\sigma_{12|3} = \sigma_{12} - \sigma_{13}\sigma_{23}/\sigma_{33}. \tag{1.9}$$

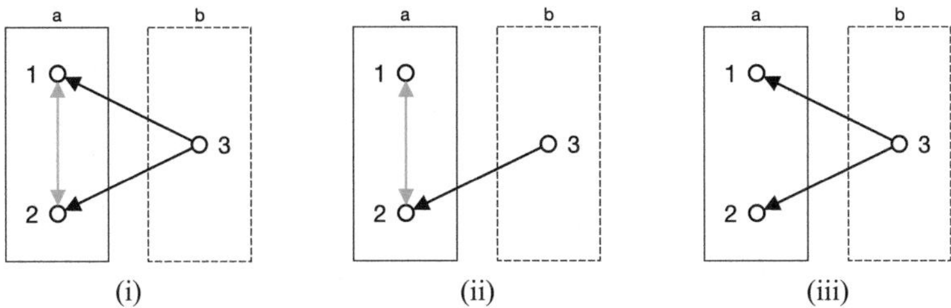

Fig. 1.4 Three regression graph models with two responses and one explanatory variable. **i** No independences. **ii** Marginal independence $X_1 \perp\!\!\!\perp X_3$. **iii** Conditional independence $X_1 \perp\!\!\!\perp X_2 \mid X_3$

More details on these equations can be found in the appendix to this chapter.

As the relevant parameters of the conditional distribution $p_{12|3}$ are the two regression coefficients and the conditional covariance we use a graph representing the variables as vertices inside the blocks a and b. Then we use a *bi-directed edge* for the conditional association $\sigma_{12|3}$ between the responses and *arrows* for the regression coefficients between a response and the explanatory variable; see Fig. 1.4, panel (i).

Notice that the block b is dashed to indicate that the variable X_3 is fixed at its observed values because we are interested only in the independence structure of the conditional distribution $p_{12|3}$.

The graph in block a can be interpreted as a *conditional covariance graph*, while the whole graph is said to be a *multivariate regression graph* because it reflects the independence structure of a multivariate regression model (1.8).

Independence models are defined naturally by constraining some of the parameters to 0. Using results for the multivariate Gaussian distribution we have that

- the constraints $\beta_{12} = 0$ or $\beta_{13} = 0$ are equivalent to the marginal independence between X_1 and X_2 or between X_2 and X_3;
- the constraint $\sigma_{12|3} = 0$ is equivalent to the *conditional independence* of X_1 and X_2 given X_3.

Figure 1.4 shows the regression graphs (ii) for the marginal independence $X_1 \perp\!\!\!\perp X_3$ and (iii) for the conditional independence $X_1 \perp\!\!\!\perp X_2 \mid X_3$. $\square$

1.3 Multivariate Regression Graphs

The conditional covariance graph of Example 1.5 has a rather limited application. Therefore, we consider now a generalization with any number of joint responses in block a and any number of explanatory variables in block b. The starting point is the factorization of the probability distribution of all the variables as

$$p_{a \cup b} = p_{a|b}\, p_b$$

in two separate models: the conditional distribution $p_{a|b}$ and the marginal distribution p_b. In the following, $p_{a\cup b}$ will be briefly referred to as p_{ab}. As a first step we can specify a multivariate regression model only for $p_{a|b}$ without modeling the joint distribution of the explanatory variables in block b. As a second step we can also model the structure of the explanatory variables.

Example 1.6 Given a joint 4-variate normal distribution we want to study now the independence structure in a situation where $X_a = (X_1, X_2)$ are joint responses and $X_b = (X_3, X_4)$ are explanatory variables. Thus, we have two blocks $a = \{1, 2\}$ and $b = \{3, 4\}$. To simplify the notation we assume that all the variables have a zero expectation, i.e., $\mu_i = 0$ for $1, \ldots, 4$. The conditional distribution of $X_a \mid X_b$ is known to be a bivariate normal with parameters

$$\boldsymbol{\mu}_{a|b} = \begin{bmatrix} E(X_1 \mid x_3, x_4) \\ E(X_2 \mid x_3, x_4) \end{bmatrix}, \quad \boldsymbol{\Sigma}_{aa|b} = \begin{bmatrix} \sigma_{11|b} & \sigma_{12|b} \\ . & \sigma_{22|b} \end{bmatrix}, \tag{1.10}$$

where the conditional expectations are linear functions of $x_b = (x_3, x_4)$

$$\begin{aligned} E(X_1 \mid x_3, x_4) &= \beta_{13} x_3 + \beta_{14} x_4 \\ E(X_2 \mid x_3, x_4) &= \beta_{23} x_3 + \beta_{24} x_4, \end{aligned} \tag{1.11}$$

and $\boldsymbol{\Sigma}_{aa|b}$ is the conditional covariance matrix of the responses given the explanatory variables. The coefficients β are partial regression coefficients. Thus, for instance, β_{13} is the regression coefficient of X_1 from X_3 adjusted for X_4 that can be denoted in Yule's notation by $\beta_{13.4}$.

Moreover, the marginal distribution of X_b is a bivariate normal with mean zero and covariance matrix

$$\boldsymbol{\Sigma}_{bb} = \begin{bmatrix} \sigma_{33} & \sigma_{34} \\ . & \sigma_{44} \end{bmatrix}.$$

Following the rules of the previous section we can draw a multivariate regression graph with two blocks and two types of edge according to the nature of the parameters: the arc $1 \leftrightarrow 2$ is a conditional covariance $\sigma_{12|b}$, while the arc $3 \leftrightarrow 4$ is a marginal covariance σ_{34}. The 4 directed edges are determined by the partial regression coefficients; see Fig. 1.5.

Fig. 1.5 Saturated multivariate regression graph with blocks $a = \{1, 2\}$ and $b = \{3, 4\}$. Correspondence between parameters and edges

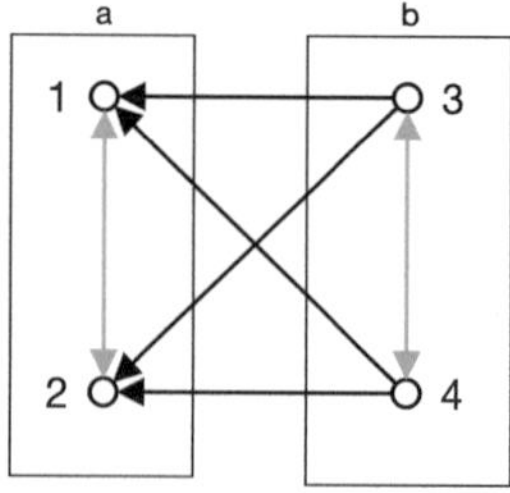

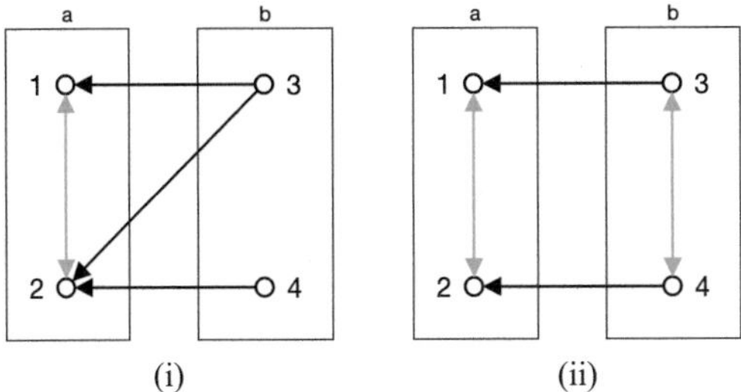

Fig. 1.6 Two multivariate regression chain graphs for the joint distribution p_{1234}. **i** Model with independencies $X_1 \perp\!\!\!\perp X_4 \mid X_3$ and $X_3 \perp\!\!\!\perp X_4$. **ii** Model with independencies $X_1 \perp\!\!\!\perp X_4 \mid X_3$ and $X_2 \perp\!\!\!\perp X_3 \mid X_4$

By the properties of the Gaussian distribution, a zero constraint for any of the parameters of Fig. 1.5 implies a removal of the associated edge and a specific conditional or marginal independence. For instance, in Fig. 1.6ii the constraint $\beta_{14} = 0$ implies the independence $X_1 \perp\!\!\!\perp X_4 \mid X_3$ and thus a missing arrow $1 \leftarrow 4$. This happens because in a Gaussian distribution the condition $E(X_1 \mid x_3, x_4) = \beta_{13}x_3$ is equivalent to $p_{1|34} = p_{1|3}$.

The further constraint $\beta_{23} = 0$ implies also a missing arrow $2 \leftarrow 3$ and the conditional independence $X_2 \perp\!\!\!\perp X_3 \mid X_4$. Finally, a zero conditional covariance $\sigma_{12|b}$ would imply a conditional independence $X_1 \perp\!\!\!\perp X_2 \mid (X_3, X_4)$ and the vanishing of the bi-directed edge $1 \leftrightarrow 2$. $\qquad\qquad\qquad\qquad\qquad\qquad\qquad\qquad\qquad\qquad\quad\Box$

The regression model introduced in previous example for the graph (ii) of Fig. 1.6 is called a *seemingly unrelated regression model*; see Zellner (1962). It can be fitted by maximum likelihood and the next example shows an application to data coming from female patients undergoing an operation.

There is an alternative specification of a multivariate regression model using the familiar linear regression notation with the error terms, as follows:

$$
\begin{aligned}
X_1 &= \beta_{13}X_3 + \beta_{14}X_4 + \epsilon_1 \\
X_2 &= \beta_{23}X_3 + \beta_{24}X_4 + \epsilon_2
\end{aligned}
\qquad
\begin{bmatrix} \epsilon_1 \\ \epsilon_2 \end{bmatrix} \overset{\text{ind}}{\sim} N\left(\begin{bmatrix} 0 \\ 0 \end{bmatrix}, \begin{bmatrix} \sigma_{11|b} & \sigma_{12|b} \\ \sigma_{12|b} & \sigma_{22|b} \end{bmatrix} \right).
$$

Sometimes the model is written compactly by using *model formulae* using the notation by Wilkinson and Rogers (1973):

$$
\begin{aligned}
X_1 &\sim X_3 + X_4 \\
X_2 &\sim X_3 + X_4 \\
(X_1, X_2) &\sim 1.
\end{aligned}
$$

Example 1.7 (Blood pressure, body mass and age) We take the data for this example from Wermuth and Cox (1995). On 44 healthy female patients expecting cosmetic surgery were collected data about diastolic and systolic blood pressure, height (cm),

Table 1.3 Estimates for a saturated multivariate regression model

Edge	Estimate	se	est/se
$1 \leftarrow 3$	-0.0042	0.0029	-1.483
$1 \leftarrow 4$	0.0003	0.0016	0.196
$2 \leftarrow 3$	0.0009	0.0037	0.244
$2 \leftarrow 4$	0.0060	0.0021	2.940
$1 \leftrightarrow 2$	-0.0055	0.0017	-3.230

weight (kg) and age (years). The variables are further transformed into the following ones:

$X_1 = \log(\text{syst/diast})$ blood pressure.
$X_2 = \log(\text{diast})$ blood pressure.
$X_3 = \text{weight}/(100 \text{ height})$ (similar to the body mass index).
$X_4 = \text{age}$.

The variables X_1 and X_2 concerning blood pressure can be considered as two responses on an equal standing (in box a) depending on body mass index and age (in box b). We assume that the data are a random sample from a bivariate Gaussian distribution for the two responses with linear conditional expectations depending both on X_3 and X_4 represented symbolically by

$$X_1 \sim X_3 + X_4, \quad X_2 \sim X_3 + X_4.$$

This model is saturated because we did not specify any constraint and the fit can be obtained by ordinary least-squares. The estimates of the parameters associated to the edges of the graph are shown in Table 1.3. The table suggests the presence of at least two conditional independencies due to the non-significant coefficients for the arrows $1 \leftarrow 4$ and $2 \leftarrow 3$. Fitting by maximum likelihood a reduced seemingly unrelated regression model defined by

$$X_1 \sim X_3, \quad X_2 \sim X_4$$

the likelihood ratio statistic is 0.075 on 2 degrees of freedom giving an indication that the data are compatible with the two conditional independencies $X_1 \perp\!\!\!\perp X_4 \mid X_2$ and $X_2 \perp\!\!\!\perp X_3 \mid X_4$ of the graph of Fig. 1.6ii. A conclusion from Table 1.3 is that diastolic blood pressure increases with age controlling for body mass, while the ratio of systolic to diastolic blood pressure is higher the lower the body mass adjusting for age. $\qquad\qquad\square$

A special case of multivariate regression graph model happens when there is a single response. This is the quite common applied situation of multiple regression. We can call the related graphs, *univariate regression graphs*.

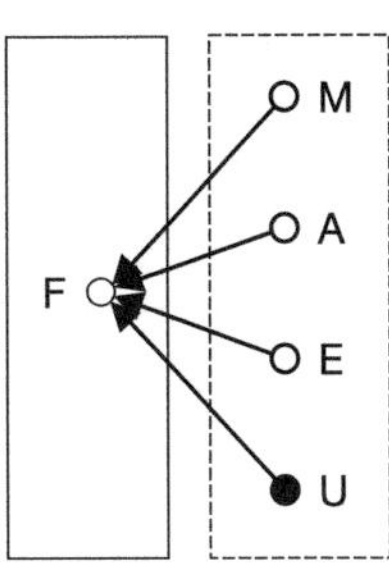

Fig. 1.7 A univariate regression graph for the dependence of fertility F on age at first marriage M, education E, age A and residence U

Example 1.8 (Fiji Fertility Survey) In the 1974 Fiji Fertility survey, one objective of the study was to assess the effects of Age A, Age at first marriage M, Education E (in completed years), Residence U (urban = 1 or rural = 0), and Fertility F (number of children ever born), given a sample of 774 married women of Fijian ethnicity, aged 35–49. Retherford and Choe (1993) fitted a linear regression model for fertility, including all the covariates and the factor Residence. The resulting model has an associate univariate regression graph shown in Fig. 1.7 where the vertices of the continuous variables are denoted by empty circles and that of the binary variable by a filled circle. In the graph of Fig. 1.7 the variables in the last block are considered as fixed at their observed values and included in a dashed box.

Under the assumptions that conditionally on all the values of the explanatory variables, the observed fertilities X_F are independently normally distributed with a mean following an additive linear model and a constant variance, the regression coefficients represent the linear dependencies of response on the explanatory variables. $\square$

1.4 Sequences of Regressions

In some research situations the variables can be grouped in more than two ordered blocks. This happens typically for cohort studies but also for cross-sectional data. In such cases the variables can be organized as in Fig. 1.8 where, for instance, there is an ordered sequence of three pairwise disjoint blocks a, b and c according to the nature of the variables. The variables in the first block a are *responses* of primary interest that can be potentially explained by the variables in b and c. The variables in the last block c are purely explanatory variables called *background variables*, or *context variables*. Finally, the variables in block b are *intermediate variables* that are explanatory for the primary responses but also responses for the context variables. The scheme is not limited to three blocks, but there can be more blocks of intermediate variables.

This set-up implies that the random vector X of all the variables is partitioned in three classes (X_a, X_b, X_c) where a, b and c are pairwise disjoint blocks of indices.

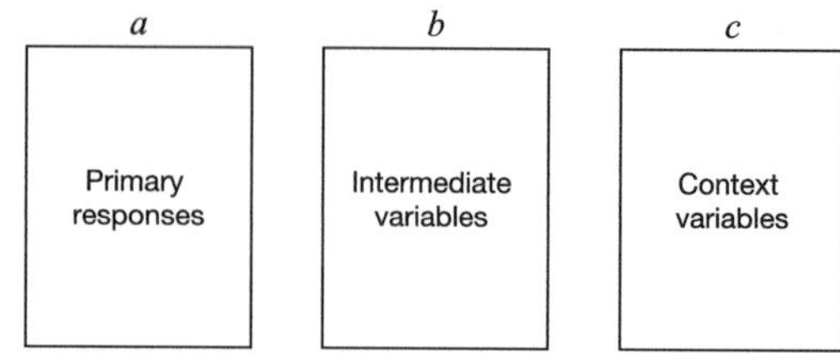

Fig. 1.8 Set-up for sequences of regressions in 3 blocks *a*, *b* and *c*

The order of the blocks is usually determined a priori by substantive subject-matter sources. Sometimes the order could be just partial because not every pair of blocks needs to be comparable.

Example 1.9 (Fiji Fertility Survey) Returning to the data from the Fiji Fertility Survey, it is sensible to create three blocks: the first containing the response Fertility, the second containing the intermediate variables Age at first marriage M and Education E and a last block with the background variables, Age A and Residence U, as shown in Fig. 1.9. This order suggests to fit regression models *recursively* using the factorization of the joint distribution

$$p_{abc} = p_{a|bc}\, p_{b|c}\, p_c. \tag{1.12}$$

The first two factors can be modeled as a univariate linear regression and a bivariate linear regression respectively:

$$
\begin{aligned}
X_F &= \beta_{FM} X_M + \beta_{FE} X_E + \beta_{FA} X_A + \beta_{FU} X_U + \epsilon_F, & \epsilon_F &\overset{\text{ind}}{\sim} N(0, \sigma_{FF|bc}) \\
X_M &= \beta_{MA} X_A + \beta_{MU} X_U + \epsilon_M, & \begin{bmatrix} \epsilon_M \\ \epsilon_E \end{bmatrix} &\overset{\text{ind}}{\sim} N\left(\begin{bmatrix} 0 \\ 0 \end{bmatrix}, \begin{bmatrix} \sigma_{MM|c} & \sigma_{ME|c} \\ \sigma_{ME|c} & \sigma_{EE|c} \end{bmatrix} \right) \\
X_E &= \beta_{EA} X_A + \beta_{EU} X_U + \epsilon_E
\end{aligned}
\tag{1.13}
$$

with an associated regression graph shown in Fig. 1.9. For simplicity, we have assumed that all the intercepts are zero.

In this case the model formulae are

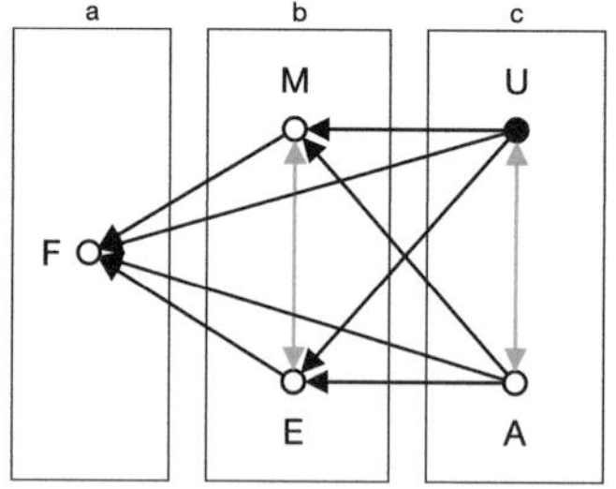

Fig. 1.9 The variables from the Fiji Fertility Survey partitioned in three blocks with a full set of directed and bi-directed edges

$$(F \sim M + E + A + U)$$
$$(M \sim U + A), \quad (E \sim U + A), \quad (M * E \sim 1)$$

where, for simplicity, we used the names instead of the symbols of the variables. $\square$

In a factorizations like that of Eq. (1.12) the terms are conditional probability distributions with specific independence structures. Each term can be interpreted as a univariate or multivariate conditional regression graph and the final term associated to the context variables can be modeled, but not necessarily, as a covariance graph.

The graph for the joint distribution of p_{abc} is called a *regression chain graph* and the blocks are called *chain components*. In general, we will denote the chain components by g_j, $j = 1, \ldots, J$ where g_1 contains the pure responses and g_J contains the context variables. The edges of the graph must respect the following rules:

1. within blocks bi-directed edges only are allowed, and no directed edges;
2. between blocks only directed edges respecting the chosen order are allowed, pointing from g_j to g_k where $j < k$.

Sometimes we use the term chain components to denote both the set of vertices in a block or the bi-directed subgraph induced by them.

Example 1.10 (**Fiji Fertility Survey: fitted models**) The saturated regression chain graph model (1.13) can be fitted by maximum likelihood with estimates and standard errors displayed in Table 1.4. Of the eight arrows associated to the parameters β three are judged to be not important leading to the elimination of the arrows $F \leftarrow E$, $E \leftarrow U$ and $M \leftarrow A$. Furthermore it turns out that the test of independence of age and residence is not rejected leading to the elimination of the bi-directed edge $A \leftrightarrow U$. The reduced model obtained after removing the two mentioned edges gives an acceptable fit with a deviance of 3.39 on 4 degrees of freedom. The final regression chain graph is shown in Fig. 1.10.

The graph can be interpreted by looking at the missing edges as specific conditional or marginal independencies as follows.

Table 1.4 Maximum likelihood estimates of the parameters of model (1.13). The labels (n.s.) indicate the non significant test statistics

Parameter	Estimate	se	z	Parameter	Estimate	se	z
β_{FM}	-0.206	0.024	-8.568	$\sigma_{FF\|bc}$	8.055	0.410	19.660
β_{FA}	0.111	0.025	4.399	$\sigma_{MM\|c}$	18.141	0.923	19.660
β_{FU}	-1.389	0.231	-6.012	$\sigma_{ME\|c}$	0.809	0.359	2.254
β_{FE}	0.009	0.044	0.211 (n.s)	$\sigma_{EE\|c}$	5.452	0.277	19.660
β_{MU}	-1.007	0.344	-2.930	σ_{AA}	16.874	0.858	19.660
β_{MA}	0.007	0.037	0.195 (n.s.)	σ_{AU}	0.005	0.066	0.072 (n.s.)
β_{EA}	-0.098	0.020	-4.818	σ_{UU}	0.199	0.010	19.660
β_{EU}	0.343	0.188	1.819 (n.s.)				

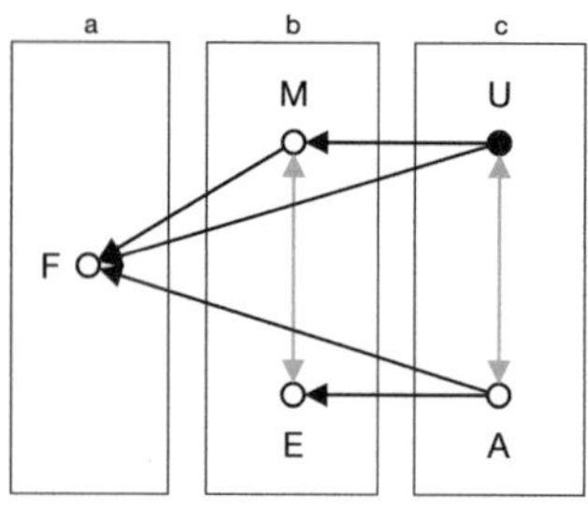

Fig. 1.10 A well-fitting regression chain graph to the data of the Fiji Fertility Survey

- The missing edge $F \leftarrow E$ implies a conditional independence $X_F \perp\!\!\!\perp X_E$ given all the other regressors in the previous blocks, that is given X_M, X_A, X_U. This means that fertility for Fijian women is independent of education given age at first marriage, age and residence.
- The missing edge $M \leftarrow A$ implies a conditional independence $X_M \perp\!\!\!\perp X_A$ given all the other regressors in the previous block, that is X_U. Thus age at first marriage and age are independent given residence.
- The missing edge $E \leftarrow U$ implies a conditional independence $X_E \perp\!\!\!\perp X_U$ given X_A. Education is independent of residence given age.
- The missing bi-directed edge $A \leftrightarrow U$ implies a marginal independence $X_A \perp\!\!\!\perp X_U$ because there are no previous blocks. Thus age and residence are independent. $\qquad\qquad\square$

1.5 The Regression Chain Graph Markov Property

A regression chain graph model is based on a specific order of the chain components $g_1, \ldots, g_J$ of the vertices associated to the variables $X = (X_1, \ldots, X_d)$. The order itself implies a *basic density factorization*

$$p_V = \prod_{j=1}^{J-1} p_{g_j \mid g_{>j}} \times p_J \tag{1.14}$$

where $g_{>j}$ denotes the *predecessors* of a given chain component g_j, that is $g_{>j} = g_{j+1} \cup \cdots \cup g_J$.

The possible constraints of independence depend on the structure of the arrows and of the bi-directed edges in the graph. We require a specific *regression chain graph Markov property*, to translate the properties of the graph into conditional independence statements; see Marchetti and Lupparelli (2011). Two concepts are important to understand this Markov property:

- the set of all the disconnected subgraphs within a chain component;
- the set of *parents* of the connected subgraphs $\mathcal{G}_A$ within a chain component, denoted by pa(A), containing all the vertices w such that $v \leftarrow w$ with $v \in A$ and w within the predecessors of a chain component.

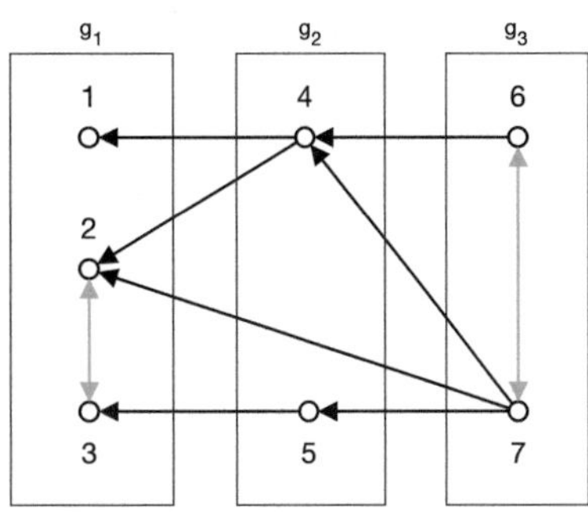

Fig. 1.11 A sequence of multivariate regression graphs with three blocks a, b and c

Example 1.11 In the regression chain graph of Fig. 1.11 there are three chain components $g_1 = \{1, 2, 3\}$, $g_2 = \{4, 5\}$ and $g_3 = \{6, 7\}$. The predecessors of g_1 are $g_1 \cup g_2 = \{4, 5, 6, 7\}$, the parents of $\{1, 2\}$ are $\{4, 7\}$. The connected components of $\{1, 2, 3\}$ are $\{\{1\}, \{2, 3\}\}$. $\square$

Definition 1.2 (*Regression chain graph (RCG) Markov property*) Let $\mathcal{G}$ be a regression chain graph with chain components g_j, $j = 1, \ldots, J$ and vertex set V. Then, the joint probability distribution p_V of a random vector X_V satisfies the *regression chain Markov property* with respect to $\mathcal{G}$ if and only if

(1) $p_{A|g_{>j}} = p_{A_1|g_{>j}} \times \cdots \times p_{A_r|g_{>j}}$ for all disconnected subgraph of g_j with connected components $A_1, \ldots, A_r$ and for all $j = 1, \ldots, J$;

(2) $p_{A|g_{>j}} = p_{A|\mathrm{pa}(A)}$ for all connected subgraphs A of g_j and for all $j = 1, \ldots, J$.

The first condition implies that if in a block g_j there is a disconnected set A with connected components $A_1, \ldots, A_r$ then there is a joint independence $X_{A_1} \perp\!\!\!\perp \cdots \perp\!\!\!\perp X_{g_{>j}}$ conditional on the predecessors. If $j = J$ is the last block then $X_{A_1} \perp\!\!\!\perp \cdots \perp\!\!\!\perp X_{A_r}$ marginally.

The second condition implies that if i and k are two vertices in different blocks g_j, $g_{>j}$ respectively, and the arrow $i \leftarrow k$ is missing, then there is a conditional independence $X_i \perp\!\!\!\perp X_k \mid X_{g_{>j}\setminus\{k\}}$.

Example 1.12 (**Continuing Example** 1.11) The regression chain graph Markov property defines a simplification of the basic factorization 1.14. For instance, considering the graph of Fig. 1.11 we have

$$
\begin{aligned}
p_{1234567} &= p_{123|4567}\, p_{45|67}\, p_{67} && \text{(basic factorization)} \\
&= p_{1|4567}\, p_{23|4567}\, p_{4|67}\, p_{5|67}\, p_{67} && \text{(connected set Markov property)} \\
&= p_{1|4}\, p_{23|457}\, p_{4|67}\, p_{5|7}\, p_{67} && \text{(parents of the connected subgraphs).}
\end{aligned}
$$

Notice that each factor of the final factorization can be represented by a regression graph or a bi-directed graph as shown in Fig. 1.12. $\square$

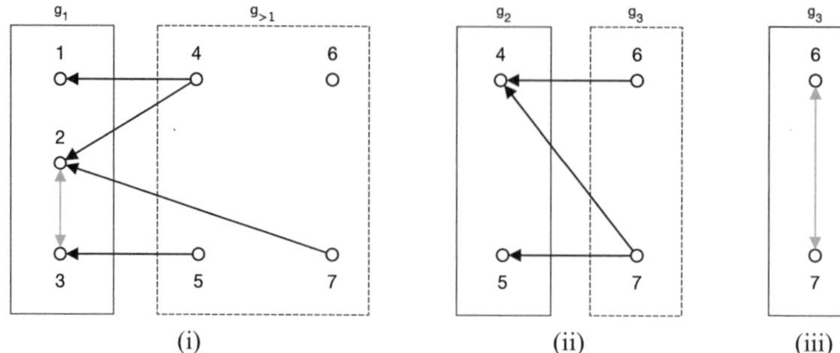

Fig. 1.12 Decomposition of the sequence of regressions of Fig. 1.11 in two regression graphs **i** for $p_{123|4567}$, **ii** for $p_{45|67}$ and **iii** a covariance graph for p_{67}

1.6 Some Applications to Categorical Data

In the previous sections we discussed only the situation where $X = (X_1, \ldots, X_d)$ is a random vector with a joint multivariate Gaussian distribution or the conditional distributions have a linear regression function $E(X_a \mid X_b)$ and a constant covariance function $\mathrm{cov}(X_a \mid X_b)$.

In such cases the regression chain graphs have a simpler interpretation in terms of linear regression equations and they can be fitted using standard methods for multivariate linear regression models. While some complications can arise if there are nonlinearities or interactions they can be solved by relatively simple methods; see Cox and Wermuth (1996). In this section we give instead just a brief introduction to models for categorical data, which will be dealt with in more detail in the next chapter. We start from a simple example.

Example 1.13 (Canadian Women's Labour-Force Participation) The data in Table 1.5 come from a social survey of the Canadian population conducted in 1977 on 263 married women between the ages of 21 and 30. The original data are extensively analyzed in Fox and Weisberg (2019). Here two of the variables were transformed to binary variables.

- L: full time work (1 = yes, 0 = no). Originally the variable had three levels.
- C: presence of children in the household (1 = yes, 0 = no).
- H: husband's income (1 = above the median of \$14,000, 0 = below the median). The original continuous variable was median dichotomized in this example.
- R: region, a factor with 5 levels `Atlantic`, Atlantic Canada; `BC`, British Columbia; `Ontario`; `Prairie`, Prairie provinces; `Quebec`.

All the variables are categorical and the data are reported in Table 1.5 as a $2 \times 2 \times 2 \times 5$ contingency table.

In this study, the response of primary interest is labour participation L which is potentially dependent on the presence of children C and husband's income H. It

Table 1.5 Contingency table for Canadian Women's Labour-Force participation. L = labour-force participation (1 = yes, 0 = no); C = presence of children in the household; H = husband's income (1 = above the median \$14, 000, 0 = below); R = region

			R, region				
L	C	H	Atlantic	BC	Ontario	Prairie	Quebec
0	0	0	1	4	6	0	5
1	0	0	1	4	11	3	10
0	1	0	11	6	25	15	18
1	1	0	2	0	4	4	4
0	0	1	1	4	6	1	5
1	0	1	1	2	10	1	3
0	1	1	11	8	44	7	19
1	1	1	2	1	2	0	1

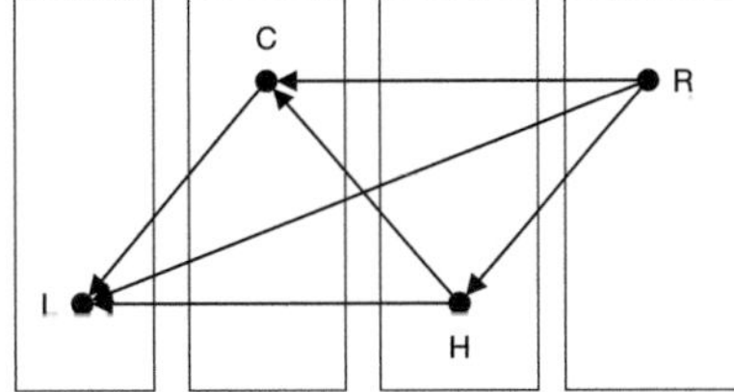

Fig. 1.13 Regression chain graph for the study on Labour-Force Participation. All variables are categorical and thus denoted by filled circles

is also reasonable to assume that the presence of children may in turn be dependent on husband's income. The region R is clearly a purely explanatory variable of non-specific nature. This ordering of the variables is displayed together with a full regression chain graph in Fig. 1.13. As all the variables are categorical, we can assume that the joint probability distribution is the *cross-classified multinomial* of size 1, defined by a list of joint probabilities π_{ijhk} indexed by $i = 0, 1$; $j = 0, 1$; $h = 0, 1$; $k = 1, \ldots, 5$. Moreover, the regression chain graph has an associated factorization

$$\pi_{ijhk} = \pi_{i|jhk}\,\pi_{j|hk}\,\pi_{h|k}\,\pi_k \tag{1.15}$$

where all the factors are conditional Bernoulli distributions. $\square$

The dependence of the conditional probabilities in a factorization like (1.15) on their respective explanatory variables can be modeled using *logistic regression*; see Cox and Snell (1989). In logistic regression the conditional probabilities π, in the range $(0, 1)$, are transformed into so-called *logits* $\Lambda(\pi) = \log\{\pi/(1 - \pi)\}$ in the range $(-\infty, +\infty)$. Then, a linear model is specified for the logits defining a linear predictor

$$\Lambda(\pi) = \sum_r x_{ir}\beta_r,$$

where the x_{ir} are the regressors. Equivalently, the probabilities can be expressed as

$$\pi = \Lambda^{-1}\left(\textstyle\sum_r x_{ir}\beta_r\right) \text{ where } \Lambda^{-1}(z) = \exp(z)/\{1 + \exp(z)\}.$$

The parameters β in the linear logistic models are variation independent and their estimates can be obtained by maximum likelihood with a standard Fisher scoring algorithm.

Example 1.14 (Canadian Women's Labour-Force Participation) Three recursive regression models specified directly by the model formulae

$$
\begin{aligned}
(L \sim C + H + R), &\quad \text{Bernoulli(link = logit)} \\
(C \sim H + R), &\quad \text{Bernoulli(link = logit)} \\
(H \sim R), &\quad \text{Bernoulli(link = logit)}
\end{aligned}
$$

are fitted using maximum likelihood. In fact, we were able to specify additive models because we previously verified the absence of interaction terms.

Then, well-fitting reduced models are considered after removing the non significant predictors. The results for the first two models are shown in Table 1.6 where the estimates, the standard errors and the deviances G^2 (the likelihood ratio test statistics) are reported.

Table 1.6 Maximum likelihood estimates of the parameters of three logistic regression models and of the related reduced models

Response F	Starting model			Reduced model, LRT = 2.81 (df = 4)		
Parameters	Estimate	se	z	Estimate	se	z
Const.	1.04	–	–	0.68	–	–
C	−2.61	0.36	−7.23	−2.44	0.33	−7.29
H	−0.77	0.35	−2.21	−0.79	0.34	−2.32
R.BC	−0.94	0.75	−1.27			
R.Ontario	−0.25	0.59	−0.43			
R.Prairie	0.17	0.69	0.24			
R.Quebec	−0.34	0.63	−0.55			

Response C	Starting model			Reduced model, LRT = 2.41 (df = 1)		
Parameters	Estimate	se	z	Estimate	se	z
Const.	1.67	–	–	1.87	–	–
H	0.44	0.28	1.55			
R.BC	−1.83	0.66	−2.78	−1.80	0.65	−2.76
R.Ontario	−1.09	0.58	−1.89	−1.05	0.58	−1.82
R.Prairie	−0.14	0.73	−0.19	−0.22	0.73	−0.31
R.Quebec	−1.25	0.60	−2.09	−1.27	0.60	−2.13

Fig. 1.14 Reduced
regression chain graph
(actually a DAG) for the data
on Canadian Women's
Labour-Force participation

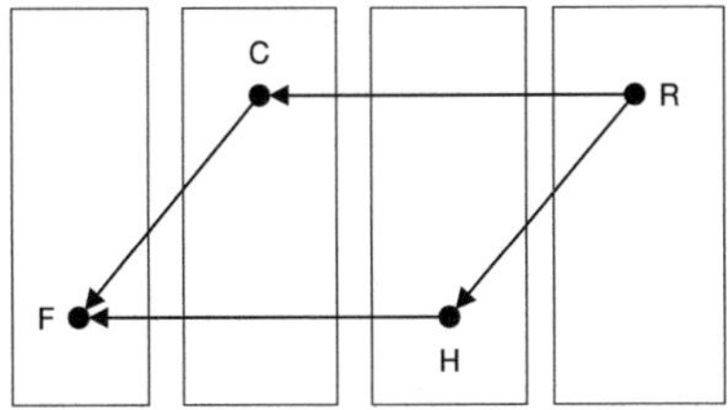

A final reduced regression graph of Fig. 1.14 is found, showing the independencies

$$X_L \perp\!\!\!\perp X_R \mid (X_C, X_H); \quad X_C \perp\!\!\!\perp X_H \mid X_R,$$

that is women labour participation is independent of region given children present
and husband income; and children present is independent of husband income given
region.   □

In the previous example we had no joint responses, and therefore we were able to
specify a regression chain graph model using a factorization and just univariate linear
logistic models. With joint responses the specification requires a more complex class
of models based on the *multivariate logistic parameterization* that will be the subject
of the next chapter. Here we give an example where the variables admit a sequence
of chain components with joint (intermediate) responses.

Example 1.15 (**U.S. General Social Survey**) Consider the following 6 variables
from the U.S. General Social Survey (Davis et al. 2007), for years 1972–2006:

D: Favor or oppose death penalty for murder (1= favor, 2 = oppose)?
A: Do you think it should be possible for a pregnant woman to obtain legal abortion
if she became pregnant as a result of rape? (1 = yes, 2 = no).
J: How satisfied are you with the work you do? (1 = very satisfied, 2 = moderately
satisfied, 3 = a little dissatisfied, 4 = very dissatisfied). Categories 3 and 4 were
merged together.
B: Confidence in banks and financial institutions (1 = a great deal, 2 = only some,
3 = hardly any).
G: Would you favor or oppose a law which would require a person to obtain a police
permit before he or she could buy a gun? (1 = favor, 2 = oppose).
S: Sex (1 = male, 2 = female).

We can interpret the intermediate responses D and G as indicators of the attitude
towards individual safety, while D and A are also indicators of the concern for the
value of human life, even in extreme situations. The basic factorization according to
the ordered blocks of the graph is

$$p_{DGABJS} = p_{D|GABJS}\, p_{GA|BJS}\, p_{BJ|S}\, p_S.$$

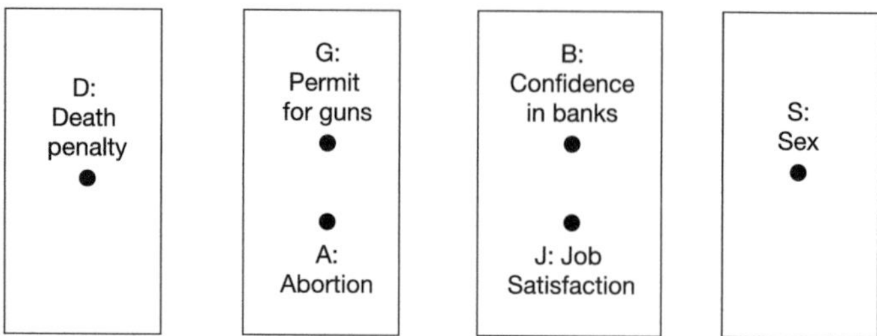

Fig. 1.15 An ordering of 6 categorical variables in 4 blocks

The fourth chain components reported in Fig. 1.15 show that, while the conditional distribution $p_{D|GABJS}$ can be modeled using a univariate linear logistic model, the conditional distribution $p_{GA|BJS}$, for instance, must be modeled using a multivariate logistic model. Essentially, this amounts to consider the vector of probabilities

$$
\begin{bmatrix}
P(X_G = 1 \mid X_B = b, X_J = j, X_S = s) \\
P(X_A = 1 \mid X_B = b, X_J = j, X_S = s) \\
P(X_G = 1, X_A = 1 \mid X_B = b, X_J = j, X_S = s)
\end{bmatrix}, \; b = 1, 2, 3; \; j = 1, 2, 3; \; s = 1, 2;
$$

and to specify a multivariate model for each component using an appropriate link function. This will be discussed in detail in Chap. 2. $\qquad\qquad\square$

1.7 Interpretation via Latent Variables

The univariate regression chain graphs like those in Figs. 1.13 and 1.14 define the subclass of directed acyclic graphs (DAG), that have only directed edges and with no directed cycles, i.e., such that there are no paths starting from a vertex, following the orientation of the edges and returning to the first vertex. When some variables associated to the vertices are *not observed* there is a general theory that explains what is the graph *induced* after marginalization over the unobserved variables. The unobserved variables are sometimes called latent variables but they are supposed to exist to understand the distribution of the observed variables.

There is an interesting case where the existence of a latent variable explains the existence of a pair of joint responses and gives rise to a multivariate regression graph structure. Assume that the DAG model of Fig. 1.16i with four observed variables $X_i, i = 1, \ldots, 4$ and one latent variable X_L, with a joint multivariate normal distribution. The regression equations are defined by

Fig. 1.16 **i** Directed acyclic graph with a latent variable L. **ii** Induced multivariate regression graph marginalizing over the latent variable L

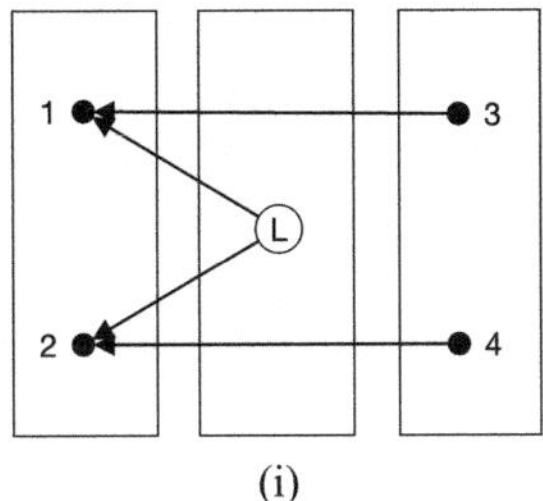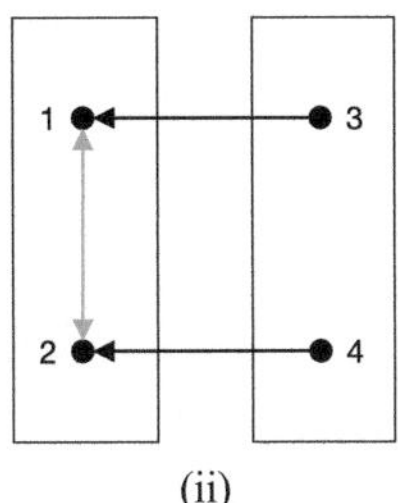

$$X_1 = \beta_{1L} X_L + \beta_{13} X_3 + \epsilon_1,$$
$$X_2 = \beta_{2L} X_L + \beta_{24} X_4 + \epsilon_2,$$
$$X_L = \epsilon_L,$$
$$X_3 = \epsilon_3,$$
$$X_4 = \epsilon_4,$$

where the errors are jointly independent with mean zero and variances δ_{ii}.

If we marginalize over the latent variable we get a normal joint distribution of the observed variables. Then it can be shown that the conditional distribution of $X_a = (X_1, X_2)$ given $X_b = (X_3, X_4)$ is a bivariate normal with expected value

$$E(X_a \mid X_b) = B X_b \quad \text{where } B = \begin{bmatrix} \beta_{13} & 0 \\ 0 & \beta_{14} \end{bmatrix}$$

and conditional covariance matrix

$$\Sigma_{aa|b} = \begin{bmatrix} \delta_{LL}\beta_{1L}^2 + \delta_{11} & \beta_{1L}\beta_{2L}\delta_{LL} \\ \cdot & \delta_{LL}\beta_{2L}^2 + \delta_{22} \end{bmatrix}$$

Therefore, the induced distribution after marginalization over X_L satisfies two conditional independencies $X_1 \perp\!\!\!\perp X_4 \mid X_3$ and $X_2 \perp\!\!\!\perp X_3 \mid X_4$ with an induced multivariate regression graph shown in Fig. 1.16ii. In this graph the bi-directed edge $1 \leftrightarrow 2$ is associated to the parameter $\sigma_{aa|12} = \beta_{1L}\beta_{2L}\delta_{LL}$ depending on the product of the two regression coefficients measuring the dependence of X_1 and X_2 from the latent variables. Therefore we can interpret the edge $1 \leftrightarrow 2$ as induced by $1 \leftarrow L \rightarrow 2$ when the latent variable is ignored.

One important feature of DAG models is that the consequences of marginalizing over latent variables can be derived form the graph itself with an explicit procedure. The algorithm is based on detecting the directed *V-configurations* defined as subgraphs of three nodes and only two directed edges. There are only three types of V-configurations in a DAG:

$$i \leftarrow \bullet \rightarrow j \qquad i \leftarrow \bullet \leftarrow j \qquad i \rightarrow \bullet \leftarrow j$$

called by Cox and Wermuth (1996) *source V*, *transition V* and *collision V*, respectively. By marginalizing over the inner node of a source in the DAG, the induced graph is obtained by removing the inner node and adding a bi-directed edge $i \leftrightarrow j$. This is the reason why the DAG in Fig. 1.16i transforms into a the regression graph of Fig. 1.16ii.

In the resulting graph the independencies of the original DAG involving the latent variable are lost, remaining only

$$
\begin{array}{lll}
X_1 \perp\!\!\!\perp X_4 & X_1 \perp\!\!\!\perp X_4 \mid X_3 & \\
X_2 \perp\!\!\!\perp X_3 & X_2 \perp\!\!\!\perp X_3 \mid X_4 & \\
X_3 \perp\!\!\!\perp X_4 & X_3 \perp\!\!\!\perp X_4 \mid X_1 & X_3 \perp\!\!\!\perp X_4 \mid X_2.
\end{array}
$$

However it can be shown that it is impossible to represent such independencies in a DAG defined on the four observed variables, but this is possible by using a regression graph as shown before in Fig. 1.16ii.

Concluding, bi-directed edges in regression graphs can be interpreted as the pairwise dependence induced after ignoring a latent variable forming a source V structure. It follows that a regression graph is always compatibile, in terms of conditional independencies, with a potential DAG including additional latent source V configurations for each bi-directed edge. This property provides useful insights on the interpretation of regression graph models and, additionaly, becomes essential to implement Bayesian inference in regression graph models for categorial data; this topic will be discussed in Chap. 4.

1.8 Appendix: Recursive Multivariate Regressions for a Gaussian Distribution

As explained in the previous sections in the Gaussian case each missing arc or arrow has an independence interpretation by means of a single vanishing parameter. Here we give the details of the parameterization for the Gaussian model represented by the graph of Fig. 1.11.

Example 1.16 (Parameterization of a Gaussian model of the Example 1.11) Assume for simplicity that the seven variables $(X_1, \ldots, X_7)$ are centered. Then the generating process can be described by the distributions

$$
X_a \mid (X_b, X_c) \sim N(\boldsymbol{B}_{a\mid b.c} X_b + \boldsymbol{B}_{a\mid c.b} X_c, \boldsymbol{\Sigma}_{aa\mid bc})
$$
$$
X_b \mid X_c \sim N(\boldsymbol{B}_{b\mid c} X_c, \boldsymbol{\Sigma}_{bb\mid c})
$$
$$
X_c \sim N(0, \boldsymbol{\Sigma}_{cc})
$$

where $\boldsymbol{B}_{a\mid b.c}$ and $\boldsymbol{B}_{a\mid c.b}$ are the matrices of regression coefficients of X_a from X_b given X_c, and of X_a from X_c given X_b, respectively and $\boldsymbol{B}_{b\mid c}$ is the matrix of regression coefficients of X_b given X_c:

$$\boldsymbol{B}_{a|b.c} = \begin{array}{c} \\ 1 \\ 2 \\ 3 \end{array}\overset{\begin{array}{cc}4 & 5\end{array}}{\begin{bmatrix} \times & 0 \\ \times & 0 \\ 0 & \times \end{bmatrix}}, \quad \boldsymbol{B}_{a|c.b} = \begin{array}{c} \\ 1 \\ 2 \\ 3 \end{array}\overset{\begin{array}{cc}6 & 7\end{array}}{\begin{bmatrix} 0 & 0 \\ 0 & \times \\ 0 & 0 \end{bmatrix}}, \quad \boldsymbol{B}_{b|c} = \begin{array}{c} \\ 4 \\ 5 \end{array}\overset{\begin{array}{cc}6 & 7\end{array}}{\begin{bmatrix} \times & \times \\ 0 & \times \end{bmatrix}}. \tag{1.16}$$

Also, $\boldsymbol{\Sigma}_{aa|bc}$, $\boldsymbol{\Sigma}_{bb|c}$ and $\boldsymbol{\Sigma}_{cc}$ are the covariance matrices of $X_a \mid (X_b, X_c)$, $X_b \mid X_c$ and X_c, respectively:

$$\boldsymbol{\Sigma}_{aa|bc} = \begin{array}{c} \\ 1 \\ 2 \\ 3 \end{array}\overset{\begin{array}{ccc}1 & 2 & 3\end{array}}{\begin{bmatrix} \times & 0 & 0 \\ 0 & \times & \times \\ 0 & \times & \times \end{bmatrix}}, \quad \boldsymbol{\Sigma}_{bb|c} = \begin{array}{c} \\ 4 \\ 5 \end{array}\overset{\begin{array}{cc}4 & 5\end{array}}{\begin{bmatrix} \times & 0 \\ 0 & \times \end{bmatrix}}, \quad \boldsymbol{\Sigma}_{cc} = \begin{array}{c} \\ 6 \\ 7 \end{array}\overset{\begin{array}{cc}6 & 7\end{array}}{\begin{bmatrix} \times & \times \\ \times & \times \end{bmatrix}}. \tag{1.17}$$

In the previous equations we used the symbol $\times$ as a substitute of a nonzero value.

Each nonzero parameter in $\boldsymbol{B}_{a|b.c}$ and $\boldsymbol{B}_{a|c.b}$ corresponds to a directed edge and each nonzero conditional covariance in $\boldsymbol{\Sigma}_{aa|bc}$ to a bi-directed edge in the graph of Fig. 1.12i. On the other hand the zeros correspond to a missing edge (directed or bi-directed) and to specific conditional independencies that can be derived from the factorization in Example 1.12. $\qquad\square$

1.9 Bibliographic Notes

Section 1.1

Conditional independence is a fundamental concept in Statistics. Basic introductions are contained in Whittaker (1990) and Wasserman (2006); a detailed mathematical treatment is Studený (2005) where are introduced the *semi-graphoid* properties of the conditional independence relation.

Section 1.2

Graphs of marginal independence for Gaussian distributions were introduced by Cox and Wermuth (1993) under the name of covariance graphs with dashed edges. The estimation of covariance graphs was based on a paper by Anderson (1973) concerning fitting linear structures in the covariance matrix.

The term covariance graph was later extended also to systems of binary or categorical variables, with graphs with bi-directed instead of dashed edges. The maximum likelihood estimation of these models was introduced by Drton and Richardson (2008) (using the iterative conditional fitting, ICF) and by Lupparelli et al. (2009) (using the marginal log-linear parameterization and a constrained optimization algorithm).

Covariance graphs are just one class of graphical Markov models for a set of variables treated on an equal standing. A much more popular class of graphical models is that of *concentration graph models* or *undirected graph models*. These graphs have full lines edges connecting the vertices and have a quite distinct interpretation with respect to covariance graphs: a missing edge $i \nsim j$ corresponds to a conditional independence between X_i and X_j given *all the remaining variables*.

The R package **ggm** by Marchetti et al. (2024) has a function to fit covariance graphs for multivariate Gaussian distributions. The fitting of conditional covariance graph can be done with the function `fitAncestralGraph`.

Section 1.3

Multivariate regression graphs are one of the types of *chain graph models* characterized by an ordered set of chain components partitioning the set of vertices. The edges within a component can be bi-directed or undirected, the edges between components are only directed but semi-directed cycles are forbidden. The original proposal of the multivariate regression graphs is in Cox and Wermuth (1993). There are several conditional independence interpretation of the chain graphs, depending on a specific Markov property. More details can be found in Drton (2009).

Section 1.4

Wermuth and Sadeghi (2012) proposed ordered sequences of univariate or multivariate regressions (i.e., regression chain graphs) as an important class of statistical models for analyzing data from randomized, possibly sequential interventions, from cohort or multi-wave panel studies, but also from cross-sectional or retrospective studies. These authors describe in detail the theory with some examples. Notice that in this general treatment of regression chain graphs the last block containing the context variables is typically modeled by an undirected graph where each missing edge between two vertices (i, j) is interpreted as a conditional independence between X_i and X_j given the remaining context variables. This rule is called the *pairwise Markov property*. In our treatment we have preferred for simplicity's sake to avoid dealing with the theory of undirected graphical models as well, by considering only simple examples with at most two context variables. For a general treatment of undirected graph models see Cox and Wermuth (1996), Lauritzen (1996) or Roverato (2017).

Section 1.5

The introduction of the multivariate regression graphs dates back to Cox and Wermuth (1993). Drton (2009) discussed a block-recursive Markov property for MRGs called *Markov property of type IV*. Definition 1.2 was introduced by Marchetti and Lupparelli (2011) and proven to be equivalent to the type IV chain graph Markov property.

Section 1.7

We have just touched on the important subject of how to represent with a suitable graph a distribution when some of the variables are not observed or are conditioned on. More specifically, we have discussed graph representations of independence models obtained by marginalizing over variables in DAGs. This task is non-trivial at all since the class of DAG models is not *stable under marginalization*; this means that the marginal independence model sometimes cannot be represented by a DAG.

Further challenges arise when the ignored variables cannot be directly observed and some assumptions are required on their probabilistic structure; this topic will be better discussed in Sect. 4.3. The class of regression graph models partially accounts for this issue because bi-directed edges represent the pairwise independence induced by marginalizing over a source node. However, also the regression graphs in general are not stable under marginalization. In this setting, it is useful to consider larger classes of graphs with additional types of edges, as summary or ancestral graphs. For a synthetic account see Wermuth (2015). The fundamental references on the subject are the papers by Richardson and Spirtes (2002), Wermuth (2011) and Sadeghi (2013).

Chapter 2
Multivariate Logistic Regression Models

Summary

In the this chapter we introduce graphical models for categorical data based on the multinomial distributions. Gaussian-based models have some similarities with multinomial models (both families are closed under the operations of marginalization and conditioning), but have also many important differences. The multinomial does not depend only on a mean vector and a covariance matrix like the Gaussian, but contains also higher order parameters called interactions. Moreover, the types of parameterization (i.e., a 1-1 transformation of the original parameters) are crucial in order to get interesting properties. One example is the logistic transformation used in Chap. 1 that will be generalized in this chapter. We introduce the multivariate logistic transformation and other generalizations called marginal parameterizations that are particularly beneficial in the context of regression graph models for categorical data. These parameterizations, based on transformations of the joint probabilities related of a contingency table, provide directly log-linear measures of association and independence constraints on the parameter space. The results concerning marginal models are illustrated through examples and case studies related to the interpretation of the regression chain graph models.

2.1 An Introduction to Marginal Models for Categorical Data

Consider data obtained by a random sample of size N from a categorical random vector $X_V = (X_v, v \in V)$, where V is the vertex set. The observed data are typically cross-classified in a multidimensional contingency table defined by the Cartesian product $\mathcal{I}_V = \mathcal{I}_1 \times \cdots \times \mathcal{I}_{|V|}$ of the levels $\mathcal{I}_v = \{1, \ldots, b_v\}$ of each variable X_v. Any cell i of the table $\mathcal{I}_V$ is a vector of the levels of the variables and $n(i)$ denotes

M. Lupparelli et al., *Regression Graph Models for Categorical Data*,
SpringerBriefs in Statistics, https://doi.org/10.1007/978-3-031-99797-6_2

the frequencies for the event $i \in \mathcal{I}_V$. The cross-classified multinomial distribution is most of the time used to model the random vector $\boldsymbol{n} = (n(\boldsymbol{i}), \boldsymbol{i} \in \mathcal{I}_V)$ of counts where the natural parameter is the joint probability vector $\boldsymbol{\pi} = (\pi(\boldsymbol{i}), \boldsymbol{i} \in \mathcal{I}_V)$ with $\mathbf{1}^\mathsf{T}\boldsymbol{\pi} = 1$. The multinomial probability model is

$$p_V(\boldsymbol{n} \mid \boldsymbol{\pi}, N) = \frac{N!}{\prod_{i \in \mathcal{I}} n(\boldsymbol{i})!} \prod_{i \in \mathcal{I}_V} \pi(\boldsymbol{i})^{n(\boldsymbol{i})}, \tag{2.1}$$

with $\pi(\boldsymbol{i}) = P(\boldsymbol{X}_V = \boldsymbol{i})$ and N is the size of the sampling scheme. The likelihood function can be also rephrased as a function of the expected cell count parameter $\boldsymbol{\mu} = N\boldsymbol{\pi}$ under the constraint $\mathbf{1}^\mathsf{T}\boldsymbol{\mu} = N$. This parameterization will prove useful in Chap. 3 to implement an algorithm for maximum likelihood estimation.

Joint probabilities are interpretable and comparable across the table, but the specification of statistical models might be difficult because the parameter space is represented by a multidimensional simplex and the models of interest often result in non-linear constraints. For instance, independence models lead to multiplicative constraints on the probabilities.

Statistical methods have been developed for binary and categorical variables to specify models of interest and to improve their interpretability. Specifically, marginal models are employed when the interest is in studying the dependencies within marginal subsets of variables based on specific measures of association aimed to describe the complexity of the dependence structure in probability tables.

The *odds ratio* (OR) represents a notable measure of association for categorical variables. It is defined by the cross-product ratio of the joint probabilities in a two-way table and quantifies the strength of dependence between two variables. In fact, we will choose parameterizations based on OR-like measures because independence conditions correspond to zero constraints on the logarithm scale. The following example illustrates the application of marginal models to specify marginal independencies.

Example 2.1 (Lienert data) Let us consider the data from Lienert (1970) which involve a random vector of three symptoms after LSD-intake recorded to be present (level 1) or absent (level 2): distortions in Affective behavior (X_1), in Thinking (X_2), and Dimming of Consciousness (X_3). A marginal model is considered to study the pairwise variable relationships.

X_1	X_2	1	2
1		25	21
2		13	14
OR$_{12}$	1.28		

X_1	X_3	1	2
1		26	20
2		15	12
OR$_{13}$	1.04		

X_2	X_3	1	2
1		23	15
2		18	17
OR$_{23}$	1.44		

The frequencies of the marginal tables show some evidence of pairwise independence since the marginal odds ratios OR$_{12} = 1.28$, OR$_{13} = 1.04$ and OR$_{23} = 1.44$ are close to one. A third-order association is considered to explore some evidence of joint independence for the trivariate distribution. The values for the conditional odds ratio between X_1 and X_2 computed in the slices of the table defined by the levels of X_3 are considerably different: 27.3 against 0.023:

X_1	X_3	1		2	
	X_2	1	2	1	2
1		21	5	4	16
2		2	13	11	1
$OR_{12\mid3}$		27.3		0.023	

Despite the pairwise independencies, the data do not support the hypothesis of mutual independence $X_1 \perp\!\!\!\perp X_2 \perp\!\!\!\perp X_3$ given the presence of a non-negligible third-order association for the variable set. The resulting marginal independence model cannot be adequately represented by a bi-directed graph which requires the connected set Markov property; see the discussion in Sect. 1.2. $\qquad\qquad\square$

The previous example shows that pairwise associations do not provide a complete picture of the dependence structure of a probability table and higher order association measures are required to identify the joint distribution. Parameterizations for multinomial models are used to provide insight on the complexity of this dependence structure. Specifically, models for categorical data founded on marginal parameterizations are used to quantify the multivariate dependence in marginal distributions.

2.2 The Multivariate Logistic Model

2.2.1 The Parameterization

The multivariate logistic model of Glonek and McCullagh (1995) represents a relevant exception within the class of marginal models and it is based on a parameterization defined by the one-to-one mapping

$$\boldsymbol{\eta} = \boldsymbol{C} \log \boldsymbol{L}\boldsymbol{\pi}, \tag{2.2}$$

where $\boldsymbol{\pi}$ is the $t \times 1$ vector of the joint probabilities with $t = \prod_{v \in V} b_v$, $\boldsymbol{L}$ is a $k \times t$ matrix such that $\boldsymbol{L}\boldsymbol{\pi}$ yields the probabilities of all the marginal tables, and $\boldsymbol{C}$ is a $(t-1) \times k$ *contrast matrix* that defines the appropriate transformations.

The resulting vector $\boldsymbol{\eta}$ is a collection of log-linear *interaction terms* $\boldsymbol{\eta}_D$ computed in any marginal table $\mathcal{I}_D$, for any subset $D \subseteq V$. If all the variables $\boldsymbol{X}_V$ are binary, the interactions are all scalars expressed by a log-linear combination of marginal probabilities

$$\eta_D = \sum_{E \subseteq D} (-1)^{|D \setminus E|} \log \pi^D(i_E), \qquad D \subseteq V \tag{2.3}$$

where $\pi^D(i_E) = P(X_D = i_E)$ is a marginal probability for the table $\mathcal{I}_D$. It is worth mentioning that the same mapping can be also defined as a log-linear combination of marginal expected cell counts $\mu^D(i_E) = N\pi^D(i_E)$ with $E \subseteq D$. The following example is illustrative for the case of two binary variables.

Example 2.2 Consider two binary variables X_1 and X_2. The multivariate logistic parameterization is defined by the following transformation:

$$
\begin{bmatrix} \eta_1 \\ \eta_2 \\ \eta_{12} \end{bmatrix} = \begin{bmatrix} -1 & 1 & 0 & 0 & 0 & 0 & 0 & 0 \\ 0 & 0 & -1 & 1 & 0 & 0 & 0 & 0 \\ 0 & 0 & 0 & 0 & 1 & -1 & -1 & 1 \end{bmatrix} \log \begin{bmatrix} 1 & 0 & 1 & 0 \\ 0 & 1 & 0 & 1 \\ 1 & 1 & 0 & 0 \\ 0 & 0 & 1 & 1 \\ 1 & 0 & 0 & 0 \\ 0 & 1 & 0 & 0 \\ 0 & 0 & 1 & 0 \\ 0 & 0 & 0 & 1 \end{bmatrix} \begin{bmatrix} \pi_{00} \\ \pi_{10} \\ \pi_{01} \\ \pi_{11} \end{bmatrix},
$$

where the π_{ij} are the joint probabilities of the variables X_1 and X_2. The main effect parameters η_1 and η_2 are the log-odds in the marginal distribution of X_1 and X_2 respectively, and the pairwise effect η_{12} is the log-odds ratio of the two-way table. The parameter $\eta_\emptyset$ is not included because it can be recovered from the sum to one constraint of the joint probabilities. $\square$

If the variables X_V are not binary, $\boldsymbol{\eta}_D$ turns out to be a vector of size $\prod_{v \in D} b_v - 1$ for any $D \subseteq V$. Basically, each parameter $\boldsymbol{\eta}_D$ is a marginal measure of association obtained as log-linear contrast of the marginal probabilities of $\mathcal{I}_D$. An analytic formula to derive the matrices C and L for any set of categorical variables is given in the Appendix 2.9 to this chapter.

Example 2.3 (Lienert data) The multivariate logistic model is here applied to Lienert data. The parameterization of the observed frequencies produces a collection of marginal parameters. The first-order parameters are the log-odds in the three marginal tables:

$$
\eta_1 = \log \frac{27}{46} = -0.53, \quad \eta_2 = \log \frac{35}{38} = -0.08, \quad \eta_3 = \log \frac{32}{41} = -0.25.
$$

The second-order parameters are the log-odds ratio of the three marginal tables:

$$
\eta_{12} = \log \frac{25 \times 14}{13 \times 21} = 0.25, \quad \eta_{13} = \log \frac{26 \times 12}{15 \times 20} = 0.04,
$$

$$
\eta_{23} = \log \frac{23 \times 17}{15 \times 18} = 0.37.
$$

The last parameter is a third-order measure of association in the full table, in particular it is the log-ratio of the conditional OR computed in a slice of the table:

$$\eta_{123} = \log \frac{1 \times 4 \times 5 \times 2}{16 \times 11 \times 21 \times 13} = -7.09.$$

The pairwise marginal independence model is supported by the fact the two-factor interaction parameters η_{12}, η_{13} and η_{23} are close to zero. $\square$

The multivariate logistic parameterization represents the marginal modeling counterpart of the well-known *log-linear parameterization* defined by the transformation

$$\boldsymbol{\theta} = \boldsymbol{M} \log \boldsymbol{\pi}, \qquad (2.4)$$

where $\boldsymbol{M}$ is a $(t-1) \times t$ matrix which makes log-linear contrasts of joint probabilities. The generic element $\boldsymbol{\theta}_D \in \boldsymbol{\theta}$ represents the D-factor log-linear interaction term, for any $D \subseteq V$. When $\boldsymbol{X}_V$ is a binary random vector the interactions are scalar quantities

$$\theta_D = \sum_{E \subseteq D} (-1)^{|D \setminus E|} \log \pi(\boldsymbol{i}_E), \qquad D \subseteq V. \qquad (2.5)$$

As before, the log-linear transformation can be defined as a log-linear combination of expected cell counts $\mu(\boldsymbol{i}_E) = N\pi(\boldsymbol{i}_E)$ with $E \subseteq D$. The difference between the log-linear and the multivariate logistic parameterizations is illustrated in the next example.

Example 2.4 For two binary variables X_1 and X_2 the log-linear parameterization is defined by

$$\begin{bmatrix} \theta_1 \\ \theta_2 \\ \theta_{12} \end{bmatrix} = \begin{bmatrix} -1 & 1 & 0 & 0 \\ -1 & 0 & 1 & 0 \\ 1 & -1 & -1 & 1 \end{bmatrix} \log \begin{bmatrix} \pi_{00} \\ \pi_{10} \\ \pi_{01} \\ \pi_{11} \end{bmatrix}.$$

The main effect parameters θ_1 and θ_2 are the logits for the conditional distribution of $X_1|\{X_2 = 0\}$ and of $X_2|\{X_1 = 0\}$, while the two-factor interaction θ_{12} is the log-odds ratio of the two-way table. Notice that $\theta_{12} = \eta_{12}$. $\square$

Example 2.5 (Lienert data) The log-linear model is now applied to Lienert's data and compared with the multivariate logistic one. The main effect parameters are conditional logits

$$\theta_1 = \log \frac{2}{21} = -2.35, \quad \theta_2 = \log \frac{5}{21} = -1.44, \quad \theta_3 = \log \frac{4}{21} = -1.65.$$

The two-factor parameters are conditional log-odds ratios:

$$\theta_{12} = \log \mathrm{OR}_{12|3} = 3.30, \quad \theta_{13} = \log \mathrm{OR}_{13|2} = 3.36, \quad \theta_{23} = \log \mathrm{OR}_{23|1} = 2.82,$$

while the three-factor log-linear interaction $\theta_{123} = -7.09$ coincides with the three-factor multivariate logistic interaction η_{123}. $\square$

Equations (2.3) and (2.5) define the same formula which is applied, respectively, to the marginal probabilities π_D to compute η_D and to the joint probabilities π to compute θ_D. Then, the multivariate logistic and the log-linear parameterization are defined by the same mapping, but applied to different probability spaces. Only the highest-order parameters coincide, i.e. $\eta_V = \theta_V$, because both are computed in the full table. It follows that the interpretation of these parameterizations is different and they are employed for different purposes in statistical analysis. In particular, the multivariate logistic parameters are pure marginal parameters while the log-linear parameters are conditional. Moreover, for the former the resulting models are *upward compatible* meaning that the parameters are *invariant with respect to marginalization* so that η_D computed in the joint probability of X_V is the same when computed in any marginal distribution of X_V including $D \subseteq V$. The same property is not satisfied in general by the log-linear parameterization.

2.2.2 Graphical Models of Marginal Independence

Models of marginal independence for categorical data can be specified by imposing zero constraints on the parameters of a multivariate logistic transformation of the joint probabilities; see Kauermann (1997).

Given two disjoint subsets of categorical variables X_A and X_B forming a contingency table $\mathcal{I}_{AB}$, the joint independence $X_A \perp\!\!\!\perp X_B$ is satisfied if and only if

$$\boldsymbol{\eta}_D = \mathbf{0}, \tag{2.6}$$

for any subset $D \subseteq A \cup B$, $D \cap A \neq \emptyset$ and $D \cap B \neq \emptyset$.

Example 2.6 (Coppen's data) Consider the data from Coppen (1966) concerning 4 psychiatric symptoms observed over a sample of 362 patients. The variables, cross-classified in Table 2.1 are: X_1: Stability (1 = extroverted, 2 = introverted), X_2: Validity (1 = psychasthenic, 2 = energetic), X_3: Depression (yes, no) and X_4: Solidity (1 = hysteric, 2 = rigid).

Table 2.1 Coppen's data concerning the variables X_1 = Stability, X_2 = Validity, X_3 = Depression and X_4 = Solidity

		X_4	1		2	
X_1	X_3	X_2	1	2	1	2
1	y		15	30	9	32
	n		25	22	46	27
2	y		23	22	14	16
	n		14	8	47	12

The chi-squared tests of the hypotheses of marginal independence $X_1 \perp\!\!\!\perp (X_3, X_4)$ and $X_4 \perp\!\!\!\perp (X_1, X_2)$, with p-values, respectively 0.14 and 0.32, are not significant and the independence model defined by the two statements gives a satisfactory fit with a deviance of 8.61 on 5 degrees of freedom. This independence model is associated with the bi-directed graph in Fig. 2.1 with the interpretation of the connected set Markov property. The set of disconnected subsets is $\mathcal{D} = \{\{1, 3\}, \{1, 4\}, \{2, 4\}, \{1, 3, 4\}, \{1, 2, 4\}\}$, and, considering the connected components of each $D \in \mathcal{D}$, the set of embedded independencies is

$$X_1 \perp\!\!\!\perp X_3, \quad X_1 \perp\!\!\!\perp X_4, \quad X_2 \perp\!\!\!\perp X_4, \quad X_1 \perp\!\!\!\perp (X_3, X_4), \quad X_4 \perp\!\!\!\perp (X_1, X_2).$$

Considering the parameterization in Eq. (2.2), the marginal independence model is given by setting

$$\eta_{13} = \eta_{14} = \eta_{24} = \eta_{134} = \eta_{124} = 0.$$

Then, the resulting marginal model includes the following non-zero terms: the main effect parameters η_1, η_2, η_3 and η_4, the marginal log-odds ratios η_{12}, η_{23} and η_{34} which measure the pairwise association detected by the bi-directed edges. The model includes also the marginal measures of association of order higher than two, η_{123}, η_{234} and η_{1234} which are included in a multinomial model, but, most of time, quite hard to interpret for categorical data. If supported by the data, the model can be simplified by including non-independence constraints on high-order parameters $\eta_{1234} = \eta_{123} = \eta_{234} = 0$ which make the independence model more parsimonious and interpretable. $\qquad\square$

In general, the multivariate logistic parameterization can be used for bi-directed graph models applied to categorical data. Lupparelli et al. (2009) proved that the multinomial probability model for a random vector $X_V = (X_i)_{i \in V}$ is Markov with respect to a bi-directed graph $\mathcal{G} = (V, E)$ if and only if

$$\boldsymbol{\eta}_D = \mathbf{0}, \quad \text{for every} \quad D \in \mathcal{D}, \tag{2.7}$$

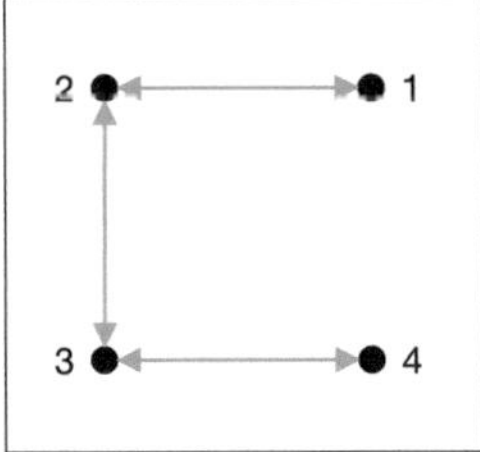
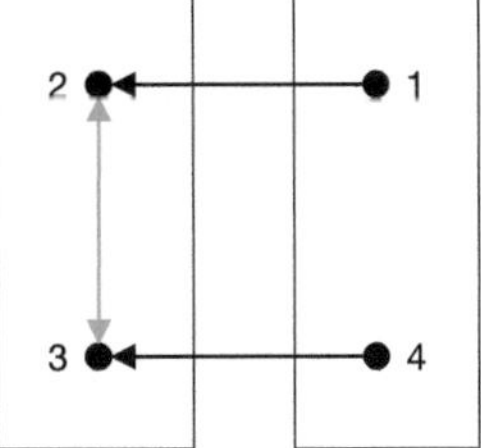

Fig. 2.1 Two Markov equivalent graphical models for Coppen's data. Left: a bi-directed graph model. Right: a regression graph model

where $\mathcal{D}$ is the set of all disconnected sets of the graph G. Basically, a bi-directed graph can be defined by setting zero constraints on the multivariate logistic interactions associated to all disconnected subsets of the graph.

Example 2.7 (Coppen's data) In Example 2.6, all symptoms are treated on the same footing. Now suppose that, based on a specific knowledge, the variables Validity and Depression are considered as joint outcomes of symptoms Stability and Solidity considered as explanatory variables. In this case a regression graph would be more appropriate to model the bivariate regression of (X_2, X_3) on (X_1, X_4). The associated graph is the regression graph shown in Fig. 2.1 to the right, which implies the basic factorization

$$p_{1234} = p_{23|14} \times p_{14}.$$

Under the RCG Markov property, the response X_2 is independent of symptom X_4 given X_1, and the response X_3 is independent of symptom X_1 given X_4, while the background variables X_1 and X_4 are marginally independent.

It can be proved that the resulting independence model $X_2 \perp\!\!\!\perp X_4|X_1$, $X_3 \perp\!\!\!\perp X_1|X_4$ and $X_1 \perp\!\!\!\perp X_4$ is equivalent to the bi-directed graph model shown in Fig. 2.1 to the left. These two graphs are said to be *Markov equivalent* because they define the same independence model. The two models nevertheless focus on different types of relationships in the variables, and, for this reason, they can be used for different interpretations in the statistical analysis. A bi-directed graph model does not impose any ordering to the variables and is useful to explore the marginal dependence structure of a multivariate distribution. When knowledge about a partial ordering of the variables is available, a regression graph model becomes more appropriate to model and interpret the direct dependencies between the variables. □

The next section discusses a class of multivariate regression models for categorical responses where the link functions are based on the multivariate logistic transformation of the conditional distribution of the response variables given the explanatory ones. This framework will be used to define the general class of the regression chain graph models.

2.3 Multivariate Regression Models for Categorical Responses

Multivariate regression models for categorical data give rise to two main challenges: (a) finding a general link function to model the non-linear dependence between the joint responses and the explanatory variables and (b) getting around the problem of the impossibility of modeling directly the association between the residuals as in the Gaussian case.

The basic idea in multivariate generalized linear models is that the residual dependence not explained by the covariate set is modeled by considering a multivariate link function; see McCullagh and Nelder (1989). Broadly speaking, a multivariate generalized linear model with categorical responses X_A and regressors X_B is defined by

$$g(\pi_{A|B}) = Z_B\beta, \tag{2.8}$$

where $\pi_{A|B} = E(X_A|X_B)$, β is a vector of regression coefficient, Z_B is the design matrix, typically block diagonal, and $g(\cdot)$ is a suitable *multivariate link function* assumed to make the relationship between the responses and the predictors $Z_B\beta$ linear. The multivariate logistic regression model, introduced by Glonek and McCullagh (1995) is a special case when the link function is $g(\pi_{A|B}) = \eta_{A|B}$, so that

$$\eta_{A|B} = Z_B\beta. \tag{2.9}$$

The following example discusses the interpretation in the specific case of Coppen's data.

Example 2.8 (**Coppen's data**) Suppose we want to regress the variables $\{X_2, X_3\}$ on the symptoms (X_1, X_4) taking values in $\{0, 1\}^2$. The equation of the multivariate logistic regression model Eq. (2.9) in this case is

$$
\begin{bmatrix}
\eta_{2|14}(0,0) \\
\eta_{2|14}(1,0) \\
\eta_{2|14}(0,1) \\
\eta_{2|14}(1,1) \\
\eta_{3|14}(0,0) \\
\eta_{3|14}(1,0) \\
\eta_{3|14}(0,1) \\
\eta_{3|14}(1,1) \\
\eta_{23|14}(0,0) \\
\eta_{23|14}(1,0) \\
\eta_{23|14}(0,1) \\
\eta_{23|14}(1,1)
\end{bmatrix}
=
\begin{bmatrix}
1 & 0 & 0 & 0 & 0 & 0 & 0 & 0 & 0 & 0 & 0 & 0 \\
1 & 1 & 0 & 0 & 0 & 0 & 0 & 0 & 0 & 0 & 0 & 0 \\
1 & 0 & 1 & 0 & 0 & 0 & 0 & 0 & 0 & 0 & 0 & 0 \\
1 & 1 & 1 & 1 & 0 & 0 & 0 & 0 & 0 & 0 & 0 & 0 \\
0 & 0 & 0 & 0 & 1 & 0 & 0 & 0 & 0 & 0 & 0 & 0 \\
0 & 0 & 0 & 0 & 1 & 1 & 0 & 0 & 0 & 0 & 0 & 0 \\
0 & 0 & 0 & 0 & 1 & 0 & 1 & 0 & 0 & 0 & 0 & 0 \\
0 & 0 & 0 & 0 & 1 & 1 & 1 & 1 & 0 & 0 & 0 & 0 \\
0 & 0 & 0 & 0 & 0 & 0 & 0 & 0 & 1 & 0 & 0 & 0 \\
0 & 0 & 0 & 0 & 0 & 0 & 0 & 0 & 1 & 1 & 0 & 0 \\
0 & 0 & 0 & 0 & 0 & 0 & 0 & 0 & 1 & 0 & 1 & 0 \\
0 & 0 & 0 & 0 & 0 & 0 & 0 & 0 & 1 & 1 & 1 & 1
\end{bmatrix}
\begin{bmatrix}
\beta_2 \\
\beta_{2|1.4} \\
\beta_{2|4.1} \\
\beta_{2|14} \\
\beta_3 \\
\beta_{3|1.4} \\
\beta_{3|4.1} \\
\beta_{3|14} \\
\beta_{23} \\
\beta_{23|1.4} \\
\beta_{23|4.1} \\
\beta_{23|14}
\end{bmatrix}. \tag{2.10}
$$

The left-hand side vector $\eta_{A|B}$ includes three sub-vectors $\eta_{2|14}$, $\eta_{3|14}$ and $\eta_{23|14}$ related to the multivariate logistic parameters for the conditional distribution of (X_2, X_3) given (X_1, X_4). The right-hand side vector β collects all the regression coefficients divided in 3 blocks that contain the regression coefficients for the univariate responses X_2, X_3 and for the bivariate response (X_2, X_3). For instance, β_2 is the intercept of the regression of X_2 on X_1, X_4 corresponding to the baseline level $x_1 = 0$, $x_4 = 0$. Furthermore, $\beta_{2|1.4}$ and $\beta_{2|4.1}$ are the partial regression coefficients of X_2 given X_1 and X_4, and $\beta_{2|14}$ is the interaction effect. Finally, the last block of regression coefficients concerns the joint responses.

Thus, the multivariate logistic link function defines a sequence of three marginal regression equations:

$$\eta_{2|14}(x_1, x_4) = \beta_2 + \beta_{2|1.4}x_1 + \beta_{2|4.1}x_4 + \beta_{2|14}x_1x_4 \tag{2.11}$$

$$\eta_{3|14}(x_1, x_4) = \beta_3 + \beta_{3|1.4}x_1 + \beta_{3|4.1}x_4 + \beta_{3|14}x_1x_4 \tag{2.12}$$

$$\eta_{23|14}(x_1, x_4) = \beta_{23} + \beta_{23|1.4}x_1 + \beta_{23|4.1}x_4 + \beta_{23|14}x_1x_4. \tag{2.13}$$

Since $\eta_{2|14}(x_1, x_4)$ and $\eta_{3|14}(x_1, x_4)$ are log-odds, Eqs. (2.11) and (2.12) define two univariate logistic regression models of X_2 and X_3, respectively, on $\{X_1, X_4\}$. Equation (2.13) is considered to model the residual association $\eta_{23|14}(x_1, x_4)$, i.e., the log-odds ratio $\mathrm{OR}_{23|14}$, between the response variables. $\qquad\square$

The multivariate logistic regression model provides a framework of regression equations for any marginal distribution of the response set. In fact, the regression of X_A on X_B in Eq. (2.9) can be written in a less compact form as a sequence of marginal regression equations

$$\eta_{D|B}(i_B) = \sum_{E \subseteq B} \beta_{D|E.\mathrm{rest}}(i_E), \qquad D \subseteq A, \; i_E \in \mathcal{I}_B. \tag{2.14}$$

When both X_A and X_B are binary, $\eta_{D|B}(i_B)$ and $\beta_{D|E.\mathrm{rest}}(i_E)$ are scalar, for any $D \subseteq A$ and $E \subseteq B$.

The interpretation of regression coefficients in a multivariate settings is not straightforward for $|D| > 1$, because $\beta_{D|E.\mathrm{rest}}(i_E)$ models the effect of X_E on η_D which is a measure of association for the marginal set X_D of response variables. To improve the interpretability of the model, further multivariate link functions and alternative model specifications could be considered. However, the multivariate logistic regression model has the advantage of being linked to the log-linear model in the joint distribution. The following example illustrates this connection and further technical details are given in Sect. 2.5.

Example 2.9 (Coppen's data) It is well-known that the coefficients of a univariate logistic regression model are identical to log-linear parameters in the joint distribution; see Whittaker (1990). For instance, in Eq. (2.11),

$$\beta_{2|1.4} = \log \mathrm{OR}_{12|4}, \qquad \beta_{2|4.1} = \log \mathrm{OR}_{24|1}, \qquad \beta_{2|14} = \log \frac{\mathrm{OR}_{12|\mathrm{rigid}}}{\mathrm{OR}_{12|\mathrm{hysteric}}},$$

where, for instance, $\beta_{2|14}$ is the third-order log-linear parameter for the table $\mathcal{I}_{124}$. Similarly, in Eq. (2.12)

$$\beta_{3|1.4} = \log \mathrm{OR}_{13|4}, \qquad \beta_{3|4.1} = \log \mathrm{OR}_{34|1}, \qquad \beta_{3|14} = \log \frac{\mathrm{OR}_{13|\mathrm{rigid}}}{\mathrm{OR}_{13|\mathrm{hysteric}}}.$$

Also, for Eq. (2.13) it can be proved that

$$\beta_{23|1.4} = \log \frac{\mathrm{OR}_{23|\mathrm{introverted, hysteric}}}{\mathrm{OR}_{23|\mathrm{extroverted, hysteric}}}, \qquad \beta_{23|4.1} = \log \frac{\mathrm{OR}_{23|\mathrm{rigid,extroverted}}}{\mathrm{OR}_{23|\mathrm{hysteric,extroverted}}},$$

and

$$\beta_{23|14} = \frac{OR_{23|\text{introverted, hysteric}} \times OR_{23|\text{extroverted, rigid}}}{OR_{23|\text{introverted,rigid}} \times OR_{23|\text{extroverted, hysteric}}}.$$

Therefore, the regression coefficients $\beta_{23|1.4}$ and $\beta_{23|4.1}$ are three-factor log-linear interactions among variables $\{X_1, X_2, X_3\}$ and $\{X_2, X_3, X_4\}$, respectively, in table $\mathcal{I}_{1234}$, and $\beta_{23|14}$ is the 4-factor log-linear interaction.   □

Regarding the *upward compatibility property* for the coefficients of a multivariate logistic regression, it occurs because it is based on a sequence of pure marginal regression equations. For instance, the regression coefficient $\beta_{2|45}$ in the multivariate regression of (X_1, X_2, X_3) on (X_4, X_5) is invariant in the regression of (X_1, X_2) on (X_4, X_5) after marginalizing over X_3. On the other hand, the regression coefficients are not invariant if some explanatory variables are omitted, except the case of special independence conditions. The next example illustrates the use of these regression models for categorical data.

Example 2.10 (GSS Data) Continuing the Example 1.15 on the GSS data we consider a smaller set of variables. The opinions D on death penalty for murder and G on Police permits for guns are considered joint responses; job satisfaction J and confidence in Banks B (both with 3 levels denoted by $(-, \cdot, +)$ with level—assumed as reference value) are considered explanatory variables.

The multivariate logistic regression model includes three equations

$$\eta_{D|BJ} = Z_{BJ}\beta_D, \quad \eta_{G|BJ} = Z_{BJ}\beta_G, \quad \eta_{DG|BJ} = Z_{BJ}\beta_{DG} \tag{2.15}$$

where $\eta_{D|BJ}$ and $\eta_{G|BJ}$ are two univariate logits and $\eta_{DG|BJ}$ is the log-odds ratio for the conditional 2×2 table of the response variables.

The matrix Z_{BJ} is a design matrix of nine columns related to the elements of the vectors β_Y, for $Y \in \{D, G, DG\}$. Each β_Y includes the intercept and four main effect parameters for the levels B_2, B_3 of the variable B and the levels J_2, J_3 of variable J and four interaction parameters $B_2 : J_2, B_3 : J_2, B_2 : J_3$ and $B_3 : J_3$. More details on the matrices Z_{BJ} are given in Table 3.2 in the next chapter.

The maximum likelihood estimates with the standard errors are shown in Table 2.2. More details about maximum likelihood inference are postponed to Chap. 3. The standard errors suggest that some interaction terms might be zero.   □

2.4 Conditional Independence Constraints

The constraints of conditional independence represented by a regression graph can be obtained by setting to zero the appropriate parameters of the multivariate logistic regression model.

Table 2.2 Maximum likelihood estimates and standard errors (se) for the saturated regression model $D \sim J * B, G \sim J * B, D * G \sim J * B$, for the reduced GSS dataset with 4 variables

Resp.	D		G		$D * G$	
Expl.	β	se	β	se	β	se
Const.	-1.023	0.05	-1.184	0.05	-0.344	0.12
J_2	0.094	0.08	0.031	0.08	0.030	0.19
J_3	0.130	0.11	-0.027	0.12	0.002	0.28
B_2	-0.045	0.06	0.008	0.06	-0.116	0.16
B_3	0.143	0.09	0.050	0.09	0.017	0.22
$J_2 : B_2$	0.030	0.10	-0.032	0.10	0.023	0.24
$J_3 : B_2$	0.258	0.13	0.001	0.15	0.108	0.34
$J_2 : B_3$	-0.071	0.13	0.183	0.14	-0.136	0.32
$J_3 : B_3$	-0.132	0.17	0.111	0.18	0.425	0.41

Example 2.11 (Coppen's data) Consider for the analysis of Coppen's data the regression graph model in Fig. 2.1 to the right. The RCG Markov property implies the independence statements $X_2 \perp\!\!\!\perp X_4 \mid X_1$ and $X_3 \perp\!\!\!\perp X_1 \mid X_4$ related to the conditional distribution $p_{23|14}$ with responses contained in the first chain component. These two independencies are satisfied by setting respectively

$$\beta_{2|4.1} = \beta_{2|14} = 0, \quad \text{and} \quad \beta_{3|1.4} = \beta_{3|14} = 0$$

in the logistic regression Eqs. (2.11) and (2.12). The marginal independence $X_1 \perp\!\!\!\perp X_4$ for the distribution p_{14} of the background variables can be satisfied by setting $\log \text{OR}_{14} = 0$. $\qquad\square$

Consider now a general regression chain graph $\mathcal{G} = (V, \mathcal{E})$ with an ordering $(g_1, g_2, \ldots, g_J)$ of the chain components and an induced factorization

$$p_V = \prod_{j=1}^{J-1} p_{g_j|g_{>j}} \times p_{g_J} \tag{2.16}$$

for the joint distribution p_V of the discrete random vector X_V.

Marchetti and Lupparelli (2011) proved that conditions (1)–(2) of the RCG Markov property are satisfied if and only if in the multivariate logistic regression equations for any $p_{g_j|g_{>j}}$ with $j = 1, \ldots, J$,

$$\eta_A(\boldsymbol{i}_{g_{>j}}) = \sum_{E \subseteq g_{>j}} \boldsymbol{\beta}_{A|E.\text{rest}}(\boldsymbol{i}_E), \quad \text{where } A \subseteq g_j, \tag{2.17}$$

so that for each $\boldsymbol{i}_E \in \mathcal{I}_{g_{>j}}$,

$$\boldsymbol{\beta}_{A|E.\text{rest}}(\boldsymbol{i}_E) = \boldsymbol{0}, \quad E \subseteq g_{>j} \backslash \text{pa}(A), \tag{2.18}$$

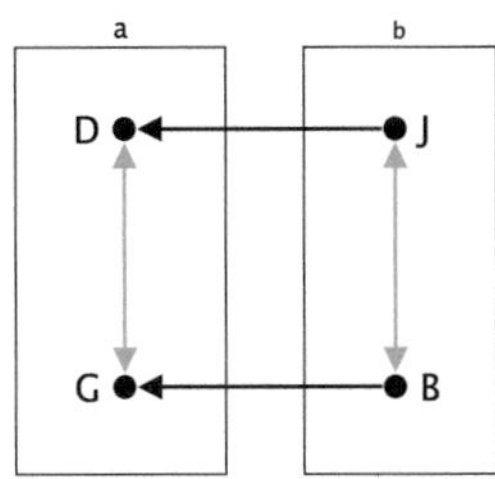

Fig. 2.2 Bivariate regression chain graph model for GSS data representing the independencies $X_D \perp\!\!\!\perp X_B | X_J$ and $X_G \perp\!\!\!\perp X_J | X_B$

if A is a connected set, and otherwise

$$\beta_{A|E.\text{rest}}(\boldsymbol{i}_E) = \boldsymbol{0}, \qquad E \subseteq g_{>j} \tag{2.19}$$

if A is a disconnected set.

Example 2.12 (GSS data) Consider the regression graph model in Fig. 2.2 for the GSS data examined in Example 2.10 where we model the dependence of the joint responses D and G in block a from the explanatory variables B and J in block b. Here it is assumed that the opinion on the gun permission is directly influenced only by the confidence in banks, while the opinion on the death penalty depends only by the job satisfaction. The responses are not independent given the explanatory factors, so that, in principle, the odds-ratio $\text{OR}_{\text{DG}|\text{BJ}}$ may depend on both covariates.

In block b the explanatory variables are marginally associated and linked by a bi-directed edge as they are considered on equal standing. When the context variables are three or more it can be more appropriate to explore conditional rather than marginal independence structures. In that case a better model choice should be an undirected graph model; see Wermuth and Sadeghi (2012).

The independencies specified by the graph of Fig. 2.2 are obtained by a reduced model with

$$\beta_{D|B.J} = \beta_{D|BJ} = \beta_{G|J.B} = \beta_{G|BJ} = \boldsymbol{0}.$$

The deviance of the fitted regression graph model is $G^2 = 14.3$ with 12 degrees of freedom. The details about the algorithm for maximum likelihood fitting of regression graph models for categorical data are postponed to Chap. 3. $\square$

Example 2.13 (GSS data) Consider regression chain graph in Fig. 2.3 for the full set of variables in the GSS data, respecting the 4 blocks introduced in Example 1.15. Given the factorization

$$p_V = p_{a|bcd} \times p_{b|cd} \times p_{c|d} \times p_d, \tag{2.20}$$

the model is decomposed in the following sequence of regressions

Fig. 2.3 Four-block
regression graph models for
GSS data

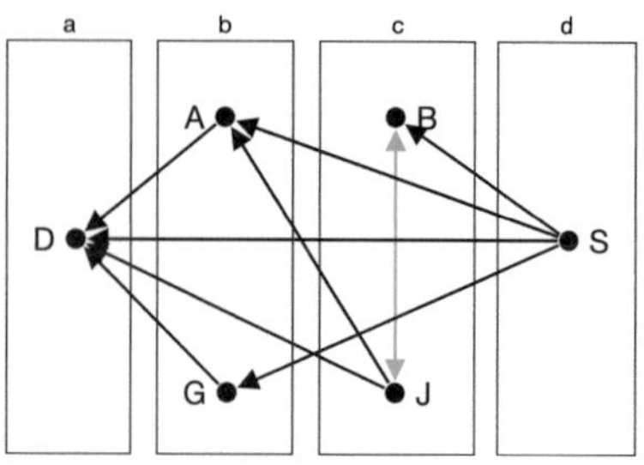

$$
\begin{aligned}
D &\sim A + G + J + S, & &\text{Bernoulli(link=logit)}\\
A &\sim J + S, \quad G \sim S, \quad A * G \sim 0 & &\text{Bernoulli(link=logit)}\\
B &\sim S, \quad J \sim 1, & &\text{Multinomial(link=multlogit)}\\
B &* J \sim S, & &\text{Multinomial(link=logOR)}
\end{aligned}
$$

On the right, the conditional distributions are specified with the specific link
functions. Notice that the different links are still produced by the multivariate logistic
parameterization. There are two models with additive effects of the regressors, but
in general it is recommended to test for the presence of interaction terms. This can
be done by comparing for instance the second model with $A \sim J * S$. The model
formula $J \sim 1$ specifies that the job satisfaction J does not depend on gender S. $\square$

Example 2.14 (GSS data) Consider now the alternative regression chain graph
shown in Fig. 2.4 with a different partition of the variables in 4 blocks and a different
dependence structure. The associated basic factorization is the same as in Eq. (2.20),
but the independencies are different:

$$
\begin{aligned}
X_D &\perp\!\!\!\perp X_B \mid (X_A, X_G, X_J, X_S), & X_A &\perp\!\!\!\perp (X_B, X_J, X_G) \mid X_S\\
(X_G, X_B) &\perp\!\!\!\perp X_J \mid X_S, & X_J &\perp\!\!\!\perp X_S.
\end{aligned}
$$

We can represent these independencies by the recursive models

Fig. 2.4 Four-block
regression graph model for
GSS data

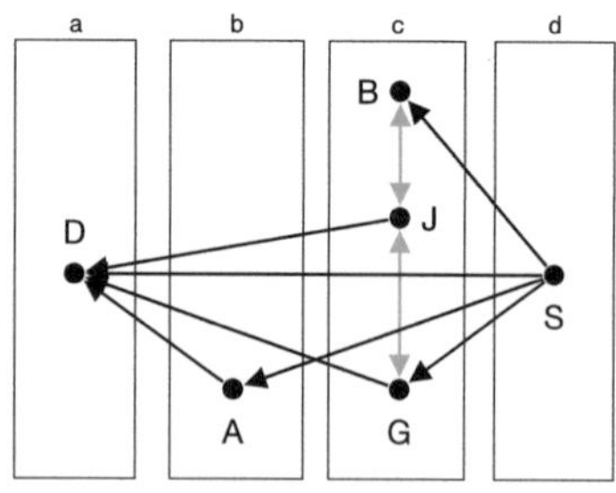

$$
\begin{aligned}
&D \sim A + J + G + S, &&\text{Bernoulli(link=logit)}\\
&A \sim S, &&\text{Bernoulli(link=logit)}\\
&G \sim S, &&\text{Bernoulli(link=logit)}\\
&B \sim S, \quad J \sim 1, &&\text{Multinomial(link=multlogit)}\\
&J * B \sim S, \quad B * G \sim 0, \quad J * G \sim S, &&\text{Multinomial(link=logOR)}\\
&B * J * G \sim S &&\text{Multinomial(link=multlogistic)}
\end{aligned}
$$

In this case the multivariate regression model for the intermediate variables in the block c must include the 2-factor and 3-factor interactions. As always, the formulae containing '~ 0' concern the zero constraints for the missing bi-directed edges. The previous model contains some non-independence constraints in the first equation that does not include interactions between explanaroty variables, but it could be also reduced by assuming that $J * G \sim 1$ or $J * B \sim 1$ or $B * J * G \sim 1$. $\square$

2.5 Log-Linear Marginal Models

As we said, the multivariate logistic parameterization is not the only modeling choice and in this section we discuss an alternative one that is based on the class of the *log-linear marginal models*. This class introduced by Bergsma and Rudas (2002) is a broad class of models for categorical variables that parameterize a joint distribution by means of the log-linear parameters related to an appropriate selection of different marginal tables. The resulting set of marginal log-linear parameters are used to model various conditional measures of association in the marginal distributions.

Any log-linear marginal model is uniquely defined by a *non-decreasing sequence* $\mathcal{M} = (M_1, \ldots, M_k, \ldots, V)$ of margins $M_k \subseteq V$ which are associated to marginal tables $\mathcal{I}_{M_k}$, such that $M_k \not\subseteq M_s$ for any $k > s$. In the following we will always assume that the sequences $\mathcal{M}$ of margins are non-decreasing. It can be seen immediately that the multivariate logistic and the log-linear parameterizations represent two extreme cases of this class when $\mathcal{M} = \mathcal{P}_0(V)$ and $\mathcal{M} = V$, respectively, with $\mathcal{P}_0(V)$ denoting the power set of V with the exclusion of the empty set.

A marginal log-linear parameterization for a categorical random vector X_V is defined by the transformation

$$
\lambda = C_{\mathcal{M}} \log L_{\mathcal{M}} \pi, \tag{2.21}
$$

where π is a $t \times 1$ vector of joint probabilities for $\mathcal{I}_V$, $L_{\mathcal{M}}$ is a $k \times t$ matrix, with $k \geq t$, which computes the probabilities of the marginal tables associated to the margins in $\mathcal{M}$ and $C_{\mathcal{M}}$ is a $(p - 1) \times k$ matrix on contrasts which defines the log-linear interaction terms in the marginal tables. Both matrices $C_{\mathcal{M}}$ and $L_{\mathcal{M}}$ are uniquely defined by the selected sequence $\mathcal{M}$ of margins. The Appendix 2.9 explains in detail the definition of the matrices $C_{\mathcal{M}}$ and $L_{\mathcal{M}}$ for any sequence $\mathcal{M}$ of margins. The generic element $\lambda_D^M \in \lambda$ represents the D-log-linear interaction term computed in the marginal table $\mathcal{I}_M$ and is defined by

$$\lambda_D^M = \sum_{E \subseteq D} (-1)^{|D \setminus E|} \log \pi^M(i_E), \quad \text{where} \quad D \subseteq M, \ M \subseteq V. \tag{2.22}$$

Example 2.15 Consider two binary variables X_1 and X_2. The marginal log-linear parameterization generated by the sequence $\mathcal{M} = (\{1\}, \{1, 2\})$ is defined by the following transformation:

$$\begin{bmatrix} \lambda_1^1 \\ \lambda_2^{12} \\ \lambda_{12}^{12} \end{bmatrix} = \begin{bmatrix} -1 & 1 & 0 & 0 & 0 & 0 \\ 0 & 0 & -1 & 0 & 1 & 0 \\ 0 & 0 & 1 & -1 & -1 & 1 \end{bmatrix} \log \begin{bmatrix} 1 & 0 & 1 & 0 \\ 0 & 1 & 0 & 1 \\ 1 & 0 & 0 & 0 \\ 0 & 1 & 0 & 0 \\ 0 & 0 & 1 & 0 \\ 0 & 0 & 0 & 1 \end{bmatrix} \begin{bmatrix} \pi_{00} \\ \pi_{10} \\ \pi_{01} \\ \pi_{11} \end{bmatrix}.$$

The main effect parameter λ_1^1 is the log-odds in the marginal distribution of X_1. The main effect parameter λ_2^{12} is the log-odds in the conditional distribution of $X_2 \mid X_1 = 0$ as well as λ_{12}^{12} is the log-odds ratio in the joint distribution of X_1 and X_2. $\qquad\square$

Equations (2.3) and (2.5) defining the multivariate logistic and the log-linear parameters η_D and θ_D, respectively, are special cases of Eq. (2.22), so that for a generic interaction indexed by D we have $\eta_D = \lambda_D^D$ and $\theta_D = \lambda_D^V$.

A key feature of marginal log-linear parameterizations is that there is a criterion to recognize whether a non-decreasing sequences of margins yields a 1-1 smooth transformation of the probabilities. Bergsma and Rudas (2002) proved a sufficient condition for a marginal log-linear parameterization to be smooth (technically a diffeomorphism) and uniquely identify the joint distribution. The sufficient condition is that the parameterization induced by $\mathcal{M}$ be *complete* and *hierarchical*.

- The parameterization is *complete* if each D-interaction term is defined in one and only one margin. This implies that the last margin in the sequence $\mathcal{M}$ must be $M_s = V$.
- The parameterization is *hierarchical* if any D-interaction term is assigned to the first margin M_k such that $D \subseteq M_k$.

The following example gives some intuitions before providing a formal definition.

Example 2.16 (**Lienert's data**) Let us consider Example 2.1 and suppose that we are interested in modeling the three bivariate marginal tables $\mathcal{I}_{12}, \mathcal{I}_{13}$ and $\mathcal{I}_{23}$ and the relative marginal log-odds ratios denoted by $\lambda_{12}^{12}, \lambda_{13}^{13}$ and λ_{23}^{23}. To get a full parameterization these three log-linear parameters can be supplemented by resorting to a complete and hierarchical marginal log-linear parameterization. For instance, given the sequence of marginal tables $\mathcal{M} = (12, 13, 23, 123)$ (omitting braces and commas for the subsets of variables) induces a complete and hierarchical parameterization

$$\lambda = (\lambda_1^{12}, \lambda_2^{12}, \lambda_{12}^{12}, \lambda_3^{13}, \lambda_{13}^{13}, \lambda_{23}^{23}, \lambda_{123}^{123})^{\mathsf{T}}$$

that include the three log-odds ratios of interest. Notice for example that in the sequence $\mathcal{M}$ the term associated to the main effect 3 is defined in the margin $\{1, 3\}$, that is the first margin including 3.

Another sequence $\mathcal{M}' = (1, 12, 23, 13, 123)$ induces a different complete and hierarchical parameterization

$$\boldsymbol{\lambda}' = (\lambda_1^1, \lambda_2^{12}, \lambda_{12}^{12}, \lambda_3^{23}, \lambda_{23}^{23}, \lambda_{13}^{13}, \lambda_{123}^{123})^\mathsf{T}$$

still including the log-odds ratios of interest while the remaining parameters are computed in different marginal tables. For instance in the sequence $\mathcal{M}'$ the main effect 3 for instance is now defined in the margin $\{2, 3\}$ instead of $\{1, 3\}$ as before. Similarly with the sequence $\mathcal{M}'$ the main effect 1 is computed in the margin $\{1\}$, while with the sequence $\mathcal{M}$ it is defined in the margin $\{1, 2\}$. $\square$

The multivariate logistic and log-linear parameterizations would have been defined, respectively, by the sequence $\mathcal{M}^\eta = (1, 2, 3, 12, 13, 23, 123)$, such that

$$\boldsymbol{\eta} = (\lambda_1^1, \lambda_2^2, \lambda_3^{23}, \lambda_{12}^{12}, \lambda_{13}^{13}, \lambda_{23}^{23}, \lambda_{123}^{123})^\mathsf{T}$$

and by the sequence $\mathcal{M}^\theta = (123)$, such that

$$\boldsymbol{\theta} = (\lambda_1^{123}, \lambda_2^{123}, \lambda_3^{123}, \lambda_{12}^{123}, \lambda_{13}^{123}, \lambda_{23}^{123}, \lambda_{123}^{123})^\mathsf{T}$$

The formal construction of a complete and hierarchical marginal log-linear parameterization from given (ordered) sequence of margins $\mathcal{M} = (M_1, \ldots, M_k, \ldots, M_s = V)$ is obtained by defining the D-interaction parameters $\boldsymbol{\lambda} = (\lambda_D^{M_k})$ for $D \in \mathcal{D}_k$, with

$$\mathcal{D}_1 = \mathcal{P}_0(M_1), \quad \mathcal{D}_k = \mathcal{P}_0(M_k) \backslash \bigcup_{s=1}^{k-1} \mathcal{M}_s, \quad \text{for } k > 1. \tag{2.23}$$

In words, the algorithm rule is that of defining within M_k all marginal log-linear parameters which have not been defined within the previous margins.

Using this rule the resulting parameterization $\boldsymbol{\pi} \mapsto \boldsymbol{\lambda}$ defined by the transformation (2.21) turns out to be a diffeomorphism, even though the elements of $\boldsymbol{\lambda}$ are not, in general, variation independent. The next section will discuss this issue in detail.

Each non-decreasing sequence $\mathcal{M}$ of margins defines a unique parameterization and any reordering of the sequence that preserves its properties yields a different parameterization. The only exception is for the ordering which generates the multivariate logistic parameterization: any complete and hierarchical re-ordering of $\mathcal{M}^\eta$ defines the same set of parameters.

2.6 Variation Independence

The examination of statistical properties associated with specific parameterizations is a valuable tool in the comparison of advantages and disadvantages regarding both modeling and inference. *Variation independence* for the vector λ is a desirable property because its parameter space is the Cartesian product of the separate ranges of its components. Therefore any set of values of the interactions is admissible and identifies a proper probability model.

The interpretation of model parameters is enhanced by this property, as the range of each measure of association remains invariant with respect to the magnitude of the other parameters. For instance, it is well-known that the log-linear parameterization for multinomial distributions is always variation independent. This represents a major merit since any vector θ of log-linear parameters for a joint probability table $\mathcal{I}$ belongs to the space $\mathbb{R}^{t-1}$, where $t = |\mathcal{I}|$. Quite the opposite, the multivariate logistic parameterization is never variation independent except for two variables. Therefore, for three or more variables, there are vectors η in $\mathbb{R}^{t-1}$ that are not compatible with any joint probability distribution π.

Moreover, variation independence is especially convenient for inference. In the Frequentist paradigm it avoids out-of-range estimates in log-likelihood maximization. and in the Bayesian paradigm it is highly recommended for the specification of priors and for the implementation of algorithms for sampling from the posterior.

Bergsma and Rudas (2002) found a necessary and sufficient condition called *ordered decomposability* for complete and hierarchical marginal log-linear parameterizations to be variation independent. A sequence of arbitrary subsets of V is said to be ordered decomposable if either it has at most two elements or the ordering of the marginals is such that for every $k = 3, \ldots, s$ the maximal[1] elements out of the margins $M_1, \ldots, M_k$, denoted say by $\{H_1, \ldots, H_\ell\}$, satisfy the so called *running intersection property*, i.e. either $\ell = 2$ or for every $h = 3, \ldots, \ell$

$$(H_1 \cup \cdots \cup H_{h-1}) \cap H_h = H_g \cap H_h, \text{ for some } g.$$

This rather technical condition can nevertheless be checked automatically with efficient algorithms to find the maximal elements and to check the running intersection property.

Example 2.17 Let $V = \{1, 2, 3, 4\}$ and consider the following three examples.

- The sequence $\mathcal{M} = (12, 1234)$ is always ordered decomposable because it includes less than three sets.
- The sequence $\mathcal{M} = (12, 134, 1234)$ induces only one subsequence $\mathcal{M}_3 = (12, 134, 1234)$ with at least three elements where the set of maximal elements is $\mathcal{H}_3 = \{1234\}$ including only the full set V and it necessarily satisfies the running intersection property.

---

[1] The maximal elements of a family of sets are those not contained in any other set of the family.

- The sequence $\mathcal{M} = (12, 23, 123, 134, 1234)$ induces three subsequences $\mathcal{M}_3 = (12, 23, 123)$, $\mathcal{M}_4 = (12, 23, 123, 134)$, $\mathcal{M}_5 = (12, 23, 123, 134, 1234)$ with related maximal elements $\mathcal{H}_3 = \{123\}$, $\mathcal{H}_4 = \{123, 134\}$ and $\mathcal{H}_5 = \{1234\}$, all of them satisfying the running intersection property.

The three marginal log-linear parameterizations generated by the above sequences are therefore variation independent.

- Consider now the sequence $\mathcal{M} = (12, 23, 134, 123, 1234)$ that is a permuted version of the last sequence. This induces three subsequences $\mathcal{M}_3 = (12, 23, 134)$, $\mathcal{M}_4 = (12, 23, 123, 134)$ and $\mathcal{M}_4 = (12, 23, 123, 134, 1234)$ with associated maximal elements $\mathcal{H}_3 = \{12, 23, 134\}$, $\mathcal{H}_4 = \{134, 123\}$ and $\mathcal{H}_5 = \{1234\}$. As the maximal element $\mathcal{H}_3$ does not satisfy the running intersection property, the parameterization is not ordered decomposable, and the variation independence does not hold. $\qquad\square$

2.7 Marginal Log-Linear Parameterization for Bi-directed Graphs

The lack of variation independence for the multivariate logistic parameterization may raise serious challenges for the inference of bi-directed and regression chain graph models, especially for the Bayesian inference which will be the subject of Chap. 4. In this section we explore variation independent parameterization for bi-directed graph models using appropriate marginal log-linear parameterizations.

Lupparelli et al. (2009) proved that a joint probability distribution of a random vector X_V of categorical variables is Markov with respect to a bi-directed graph $\mathcal{G} = (V, \mathcal{E})$, if one chooses the marginal log-linear parameterization generated by the sequence $\mathcal{M} = \{\mathcal{D}, V\}^*$ where $\mathcal{D}$ is the set of all the disconnected subgraphs and $\{\}^*$ denotes a non-decreasing ordering of the sets, with the constraints

$$\lambda_D^D = \mathbf{0}, \quad \text{for all} \quad D \in \mathcal{D}. \tag{2.24}$$

Notice that the independence constraints are specified by zero restrictions on the same set of multivariate logistic parameters defined in Eq. (2.7) because $\lambda_D^D = \eta_D$. In this marginal log-linear parameterization the nonzero interactions are arranged in different marginal tables so that, in some case, variation independence can be satisfied. The bi-directed graph model for Coppen's data is an example.

Example 2.18 (Coppen's data) The four-chain bi-directed graph of Fig. 2.1 (left) has 3 disconnected sets that can be arranged in the non-decreasing sequence $\mathcal{D} = (14, 134, 124)$. Then the marginal log-linear parameterization generated by the sequence $\mathcal{M} = (14, 134, 124, 1234)$ is ordered decomposable and therefore it is variation independent. The marginal independence model is defined by setting

$\lambda_{14}^{14} = \lambda_{13}^{134} = \lambda_{134}^{134} = \lambda_{12}^{124} = \lambda_{124}^{124} = 0$, which are indeed the multivariate logistic interactions associated to the disconnected sets of the graph. $\qquad\square$

2.8 Marginal Log-Linear Parameterizations for Regression Graphs

As mentioned in Sect. 2.3, there is an equivalence between the coefficients of a multivariate logistic regression models and the marginal log-linear parameters generated by a suitable sequence of margins. Here we give the details of the marginal log-linear parameterization for the regression chain graphs where this equivalence still holds and allows us to prove some additional properties.

Given a regression chain graph $\mathcal{G} = (V, \mathcal{E})$ with an ordering of the chain components g_j for $j = 1, \ldots, J$, its marginal log-linear parameterization is generated by a non-decreasing ordering of the margins

$$\{M \cup g_{>j} \colon M \subseteq g_j\}^*, \quad j = 1, \ldots, J \tag{2.25}$$

where M is one of the possible subsets of the block of responses g_j, and $g_{>j}$ is union of the preceding blocks. Notice that the last margin of the sequence is always the full set V.

The constraints on the parameters that define the RCG Markov property are defined as follows. Given any margin $M \cup g_{>j}$ where $M \subseteq g_j$, if M is disconnected, impose

$$\lambda_{M \cup E}^{M \cup g_{>j}} = 0, \quad \text{for all } E \subseteq g_{>j} \tag{2.26}$$

and if M is connected, impose

$$\lambda_{M \cup E}^{M \cup g_{>j}} = 0, \quad \text{for all } g_{>j} \setminus \mathrm{pa}(M) \subseteq E, \tag{2.27}$$

where $E \subseteq g_{>j}$ and $\mathrm{pa}(M)$ is the set of parent nodes of M in the associated regression graph. Note that constraints in (2.27) imply that the non-zero parameters $\lambda_{M \cup E}^{M \cup g_{>j}}$ associated to a connected set M are defined for any $E \subseteq \mathrm{pa}(M)$.

Also it can be proven that for any $M \subseteq g_j$ and $E \subseteq g_{>j}$,

$$\beta_{M|E.\mathrm{rest}} = \lambda_{M \cup E}^{M \cup g_{>j}}. \tag{2.28}$$

The above identity implies that the multivariate logistic regression coefficients, within the sequence of regressions induced by a regression graph, are marginal log-linear parameters computed in the marginal table associated with a subset of a chain component plus its predecessors. Therefore, the multivariate logistic regression parameterization for the regression graphs is equivalent to the marginal log-linear parameterization generated by the sequence of margins (2.25) under the constraints of Eqs. (2.26) and (2.27) that are in turn equivalent to the constraints of Eqs. (2.18) and (2.19), respectively.

Exploiting this equivalence we get a worthwhile result stating that a multivariate logistic parameterization for a regression graph model is variation independent if the equivalent marginal log-linear parameterization is generated by an ordered decomposable sequence of margins.

Example 2.19 (GSS data) Consider again the example of the GSS data and the regression graph in Fig. 2.3 with the sequence of margins

$$\mathcal{M} = (S, BS, JS, BJS, ABJS, GBJS, AGBJS, DAGBJS)$$

given by Eq. (2.25). It can be verified that the sequence $\mathcal{M}$ is ordered decomposable thus generating a variation independent parameterization. Thus the appropriate independence constraints are obtained from Eqs. (2.26) and (2.27). $\square$

Example 2.20 (GSS data) However, the variation independent parameterization cannot always be found. Consider the regression graph in Fig. 2.4 with a block c containing three variables concerning confidence in banks, job satisfaction and opinion on gun. The non-decreasing sequence of margins associated with the graph, based on Eq. (2.25) is

$$\mathcal{M}' = (S, BS, JS, GS, BJS, BGS, JGS, BJGS, ABJGS, DABJGS)$$

that in this case is not ordered decomposable. Therefore, there is no chance of getting a variation independent parameterization. $\square$

An interesting consequence of the equivalence discussed in this section is that a sufficient condition for the variation independence of a multivariate logistic parameterization for the regression graphs is that all the chain components include at most two variables. Indeed in this case the sequence of margins turns out to be always ordered decomposable, because there is always a single maximal element of the cumulative set of margins, that trivially satisfies the running intersection property.

2.9 Appendix: Link Functions for Marginal Log-Linear Parameterizations

In this section we explain how to define the contrast and marginalization matrices $C_\mathcal{M}$ and $L_\mathcal{M}$ needed in Eq. (2.21). Both matrices are obtained by using Kronecker products that are useful and compact, but need an efficient coding. Also we shall consider only contrasts of type *local* defining baseline coding for nominal categorical variables. Contrast for ordinal categorical variables can also be defined; see Bartolucci et al. (2007).

We assume as usual a random vector $\boldsymbol{X}_V = (X_v)_{v \in V}$ of discrete variables with levels $1, \ldots, b_v$. We index the interactions by a set D inside a margin M. Then the marginalization matrix $\boldsymbol{L}_{\mathcal{M}}$ is obtained by stacking matrices $\boldsymbol{L}_{D,M}$ one below the others respecting the order of the interactions D, where

$$\boldsymbol{L}_{D,M} = \bigotimes_{v \in V} \boldsymbol{L}_{D,M,v}, \quad \text{where } \boldsymbol{L}_{D,M,v} = \begin{cases} \begin{bmatrix} \boldsymbol{A}_{0,v} \\ \boldsymbol{A}_{1,v} \end{bmatrix} & \text{if } v \in D \\ \boldsymbol{u}_{b_v}^\mathsf{T} & \text{if } v \in M \backslash D \\ \boldsymbol{1}_{b_v}^\mathsf{T} & \text{if } v \in V \backslash M, \end{cases} \tag{2.29}$$

and where $\boldsymbol{A}_{0,v}^\mathsf{T}$ is a $b_v \times b_v$ identity matrix with the last column removed and $\boldsymbol{A}_{1,v}^\mathsf{T}$ is a $b_v \times b_v$ identity matrix with the first column removed. Moreover, $\boldsymbol{u}_{b_v}^\mathsf{T}$ is a row vector with the first element equal to 1 and the remaining $b_v - 1$ elements equal to 0 and $\boldsymbol{1}_{b_v}^\mathsf{T}$ is a row vector of ones of length b_v.

The contrast matrix $\boldsymbol{C}_{\mathcal{M}}$ is a block diagonal matrix with blocks $\boldsymbol{C}_D$ for every $D \in \mathcal{P}_V \backslash \emptyset$,

$$\boldsymbol{C}_D = \bigotimes_{v \in V} \boldsymbol{C}_{D,v}, \tag{2.30}$$

where

$$\boldsymbol{C}_{D,v} = \begin{cases} [-\boldsymbol{I}_{b_v-1} \quad \boldsymbol{I}_{b_v-1}] & \text{if } v \in D \\ 1 & \text{otherwise.} \end{cases}$$

The link function just defined holds also in special cases such as the log-linear and the multivariate logistic models, but can be simplified. Thus, the marginalization matrix for the loglinear models is absent because $\boldsymbol{L}_{\mathcal{M}}$ is just an identity matrix, while for the multivariate logistic models

$$\boldsymbol{L}_M = \bigotimes_{v \in V} \boldsymbol{L}_{M,v}, \quad \text{where} \quad \boldsymbol{L}_{M,v} = \begin{cases} \boldsymbol{I}_{b_v} & \text{if } v \in M \\ \boldsymbol{1}_{b_v}^\mathsf{T} & \text{otherwise.} \end{cases} \tag{2.31}$$

2.10 Bibliographic Notes

Section 2.1

Several marginal models have been proposed to model and compare marginal distributions of categorical variables. These models are based on different parameterizations of the joint probability vector associated to a contingency table. Fundamental references are Glonek and McCullagh (1995), Ekholm et al. (1995), Bergsma and Rudas (2002) and Roverato et al. (2013). These are mixed parameterizations obtained as log-linear combination of means and canonical parameters in the exponential family theory (Barndorff-Nielsen 1978). A review of marginal models with some applications is provided by Bergsma et al. (2009).

Section 2.2

The multivariate logistic parameterization of Glonek and McCullagh (1995) can be used to specify models of marginal independence for categorical variables. Preliminary results on this have been developed by Kauermann (1997), regardless the context of graphical models. Later, Lupparelli et al. (2009) formally proved that a bi-directed graph model for categorical data is defined by a multivariate logistic parameterization with zero constraints for the parameters associated to the disconnected subsets of a bi-directed graph. Further parameterizations have been developed based on different mean (and not mixed) parameterizations by Drton and Richardson (2008) and Roverato et al. (2013).

Section 2.3

The literature on multivariate regression models for categorical responses dates back to McCullagh and Nelder (1989) which is a basic reference for generalized linear models. The multivariate dependence in the conditional distribution of the response variables given the explanatory ones is modeled by considering multivariate regression equations where the link functions are measures of associations related to marginal distributions of the response variables. Parameterizations and inference for these models have been studied by Lang and Agresti (1994), Lang (1996), by Colombi and Forcina (2001) and Bartolucci et al. (2007) with special focus on models for categorical and ordinal response variables and by Ekholm et al. (1995) and Lupparelli and Roverato (2017) where the link functions are measures of associations defined by mean parameterizations.

Section 2.4

Marchetti and Lupparelli (2011) proposed the multivariate logistic parameterization for regression graph models with some illustrative examples and interpretation of the model parameters. Regression graph models for categorical data have been also studied by Drton (2009), by Roverato (2017) and La Rocca and Roverato (2018) using parameterizations based on the Möbius inversion. These are pure marginal parameterizations because the interaction terms are function of mean parameters that are marginal probabilities in a multinomial model. Exploiting the Möbius inversion, the inverse mapping of these parameterizations is always available in closed form, unlike the multivariate logistic transformation that needs to be numerically computed.

Section 2.5

Marginal log-linear models, introduced by Bergsma and Rudas (2002), have been widely used for the analysis of categorical data, regardless of the graphical modeling context; see Bartolucci et al. (2007), Forcina et al. (2010), Rudas et al. (2010) and Evans and Richardson (2013).

Section 2.6

This section explains the importance of variation independence in statistical models for categorical data. There is an easy-to-check condition based on the running intersection rule (Lauritzen 1996) to verify whether a marginal log-linear model for a bi-directed graph is variation independent. Technical aspects and illustrative examples are given in Bergsma and Rudas (2002) and in Lupparelli (2006). An R function `rip` to understand and check the running intersection property is contained in the package **gRbase** by Dethlefsen and Højsgaard (2005).

Section 2.7

Marginal log-linear parameterizations for bi-directed graph models have been extensively discussed and compared with other parameterizations in Lupparelli (2006) and in Marchetti and Lupparelli (2011). This approach can be adopted when variation independence for the model parameters become relevant for interpretative or inferential reasons. Ntzoufras et al. (2019) developed an approach for prior specification on this class of models with related techniques for Bayesian inference which will be better discussed in Chap. 4.

Section 2.8

The multivariate logistic parameterization of Glonek and McCullagh (1995) represents a special case in the class of marginal log-linear models of Bergsma and Rudas (2002). This feature is exploits for a twofold reason: (i) to provide an interpretation of the multivariate logistic regression coefficients and (ii) verify whether a multivariate logistic models for a regression graph is variation independent. Technical aspect about the equivalence between multivariate logistic regression coefficients and marginal log-linear parameter can be read in Marchetti and Lupparelli (2011) and in Rudas et al. (2010).

Chapter 3
Maximum Likelihood Inference

Summary

This chapter is concerned with maximum likelihood inference of discrete regression graph models. An optimization procedure is discussed based on a gradient-ascent algorithm to maximize the log-likelihood function where independence constraints are specified through Lagrange multipliers. The algorithm can be recursively applied to the sequence of multivariate regression models induced by the basic factorization of a regression graph. Model selection based on a structural learning of the DAG of the chain components is also discussed with reference to some illustrative applications.

3.1 Likelihood Function for Discrete Regression Graph Models

Consider N observations from a discrete random vector X_V following a *multinomial distribution* of size 1 with probability parameter π, and let n be the vector of observed counts obtained after cross-classifying the observed data, with $\mathbf{1}^{\mathsf{T}} n = N$, and let $\mu = E(n)$ the vector of the expected cell counts. Algorithms for maximum likelihood estimation (MLE) can be better implemented in the space of the logarithm of the expected cell counts $\omega = \log \mu$ rather than in the simplex space since parameter ω belongs to a variation independent domain and this avoids the risk of out-of-range estimates during the maximization procedure.

Given a regression graph $\mathcal{G}$ associated to a random vector X_V, considering the basic factorization (1.14) of the joint distribution for a given well-ordering of the chain components $g_1, \ldots, g_J$, the likelihood function turns out to be

$$L_V(\omega) = \prod_{j=1}^{J-1} L_j(\omega_{g_j|g_{>j}}) L_J(\omega_{g_J}) \tag{3.1}$$

© The Author(s), under exclusive license to Springer Nature Switzerland AG 2025
M. Lupparelli et al., *Regression Graph Models for Categorical Data*,
SpringerBriefs in Statistics, https://doi.org/10.1007/978-3-031-99797-6_3

where $L_j(\boldsymbol{\omega}_{g_j|g_{>j}})$ is the likelihood function for the conditional multinomial distribution of $\boldsymbol{X}_{g_j}|\boldsymbol{X}_{g>j}$ with log-expected cell count vector $\boldsymbol{\omega}_{g_j|g_{>j}}$. The log-likelihood function is defined accordingly as a sum

$$\ell_V(\boldsymbol{\omega}) = \sum_{j=1}^{J} \ell_{g_j|g_{>j}}(\boldsymbol{\omega}_{g_j|g_{>j}}). \tag{3.2}$$

A probability model $p_V(\boldsymbol{\omega})$ is Markov with respect to a regression graph $\mathcal{G}$ if the independence statements defined by the RCG Markov property are satisfied for each conditional distribution of $\boldsymbol{X}_{g_j}|\boldsymbol{X}_{g>j}$. The parameterizations introduced in Chap. 2 enable the specification of the independence model by setting zero constraints on log-linear transformations of probabilities or, equally, of log-expected counts of type

$$\boldsymbol{h}(\boldsymbol{\omega}_{g_j|g_{>j}}) = \boldsymbol{0}, \qquad j = 1, \ldots, J, \tag{3.3}$$

for a suitable function $\boldsymbol{h}(\cdot)$ of the parameter $\boldsymbol{\omega}_{g_j|g_{>j}}$.

Likelihood inference of regression graph models consists in solving a constrained maximum problem that can be carried out by maximizing the *Lagrangian log-likelihood* function

$$L_V(\boldsymbol{\omega}, \boldsymbol{\tau}) = \sum_{j=1}^{J} L_{g_j|g_{>j}}(\boldsymbol{\omega}_{g_j|g_{>j}}), \tag{3.4}$$

where

$$L_{g_j|g_{>j}}(\boldsymbol{\omega}_{g_j|g_{>j}}, \boldsymbol{\tau}) = \ell_{g_j|g_{>j}}(\boldsymbol{\omega}_{g_j|g_{>j}}) + \boldsymbol{\tau}^{\mathsf{T}}\boldsymbol{h}(\boldsymbol{\omega}_{g_j|g_{>j}}), \qquad j = 1, \ldots, J \tag{3.5}$$

and $\boldsymbol{\tau}$ is the vector of *Lagrangian multipliers* associated to the zero constraints.

Next section describes an algorithm for the Lagrangian log-likelihood maximization of a broad class of multivariate regression models for categorical data where the use of the multivariate logistic transformation as link function represents a relevant instance. MLEs for regression graph models can be derived by recursively applying this algorithm for fitting the multivariate logistic regression model of $\boldsymbol{X}_{g_j}$ on $\boldsymbol{X}_{g_{>j}}$ under independence constraints, for any $j = 1, \ldots, J$.

3.2 MLEs for a Generalized Class of Non-linear Regression Models

Given two vectors $\boldsymbol{Y}$ and $\boldsymbol{X}$ of categorical response and explanatory variables, respectively, Lang (1996) developed an algorithm inspired by Aitchison and Silvey (1958) for fitting a class of non-linear regression models defined by

$$\boldsymbol{g}(\boldsymbol{\omega}_{Y|X}) = \boldsymbol{Z}_r\boldsymbol{\beta}, \tag{3.6}$$

where $g(\cdot)$ is a multivariate link function and Z_r is the rectangular matrix for a reduced model obtained by removing selected columns of the matrix Z_X of the full model in order to define zero constraints on the vector β of regression coefficients. Then, the model constraints are given by

$$U_r^\mathsf{T} g(\omega_{Y|X}) = 0, \tag{3.7}$$

where U_r is the orthogonal complement of Z_r such that $U_r^\mathsf{T} Z_r \beta = 0$.

Thereafter, a simplified notation is adopted to denote the conditional distribution of $Y|X$ for any value of the conditioning set. Let $k = 1, \ldots, K$ denotes a generic cell i_X of the probability table $\mathcal{I}_X$ and let $\mathcal{Y} = \mathrm{vec}(Y_k)_{k=1,\ldots,K}$ be the collection of the conditional response variables given each level of the covariate set; $\mathrm{vec}(\cdot)$ is the operator used to denote a column vector of element Y_k, with $k = 1, \ldots, K$. If each $Y_k \sim \mathrm{Mult}(n_k, \pi_k)$, the vector $\mathcal{Y}$ follows a product-multinomial distribution and the log-likelihood function is

$$\ell(\omega) = n^\mathsf{T} \omega - 1^\mathsf{T} \exp(\omega), \tag{3.8}$$

with $\omega = \mathrm{vec}(\omega_k)_{k=1,\ldots,K}$ and $n = \mathrm{vec}(n_k)_{k=1,\ldots,K}$. The Lagrangian log-likelihood including the zero model constraints is

$$L(\omega, \tau) = n^\mathsf{T} \omega - 1^\mathsf{T} \exp(\omega) + \tau^\mathsf{T} h(\omega), \tag{3.9}$$

where $h(\omega) = U_r^\mathsf{T} g(\omega_{Y|X})$.

3.2.1 The Algorithm

The score vector and the Hessian matrix of (3.9) need to be computed to implement a gradient-ascent algorithm for the MLE of the parameter vector $\gamma = \mathrm{vec}(\omega, \tau)$. The score $f(\gamma)$ including the first order partial derivatives is

$$f(\gamma) = \begin{bmatrix} n - \mu + \dfrac{dh(\omega)}{d\omega} \tau \\[2ex] h(\omega) \end{bmatrix}.$$

The partial derivative $\frac{dh(\omega)}{d\omega}$ is a full column rank matrix that will be shortly denoted by $H(\omega)$. If $g(\cdot)$ is the multivariate logistic link function defined in (2.9), $H(\omega) = D(\omega) L^\mathsf{T} \Psi^{-1}(\omega) Z_r^\mathsf{T} U_r$, where $D(\omega)$ is the diagonal matrix of the expected count μ and $\Psi(\omega)$ is a diagonal matrix with elements $L\mu$ that are the expected marginal cell counts. The Hessian matrix $F(\gamma)$ of the second order partial derivatives is

$$F(\gamma) = \begin{bmatrix} -D(\omega) + \dfrac{dH(\omega)}{d\omega^{\mathsf{T}}}(\tau \otimes I_s) & H(\omega) \\[2mm] \cdot & \mathbf{0} \end{bmatrix}.$$

Hereafter, a shorthand notation is used which omits the dependence on the parameter vector ω or γ for the matrices involved in the score vector and in the Hessian, e.g., $f(\gamma) = f, h(\omega) = h, H(\omega) = H$ and $F(\gamma) = F$. The notation $f^{(i)}, F^{(i)}, \omega^{(i)}$ and $\tau^{(i)}$ is instead introduced to denote the score vector, the Hessian matrix and updated parameters at each i-iteration. A gradient-ascent iteration scheme of Fisher scoring type is

$$\gamma^{(i+1)} = \gamma^{(i)} - \text{step} \times [G^{(i)}]^{-1} f^{(i)}.$$

where $G^{(i)} = E(-F^{(i)})$ is the Fisher information matrix at i-iteration. It is worth remarking that MLEs under zero constraints are in correspondence of a saddle point of the Lagrangian log-likelihood $L(\omega, \tau; n)$, then this type of algorithms typically requires to include a step criterion to improve the convergence. Step halving is a common strategies that works quite well in these cases. However more sophisticated step criterions are available; see for instance Bergsma (1997). Another crucial aspect is computing the inverse matrix $[G^{(i)}]^{-1}$ that can be computationally demanding for large tables; Appendix 3.5 discusses two possible approximations of $G^{(k)}$ that can be used to make the algorithm more scalable.

The iterative scheme can be split to update ω and τ, respectively,

$$\omega^{(i+1)} = \omega^{(i+1)} + \text{step} \times D^{-1}(n - \mu + H\tau^{(i)}) \tag{3.10}$$

$$\tau^{(i+1)} = -P(H^{\mathsf{T}} D^{-1}(n - \mu) + h), \tag{3.11}$$

where $P = (H^{\mathsf{T}} D^{-1} H)^{-1}$.

This algorithm provides some advantages: (i) it does not require the computation of the inverse link function $g(\pi) \mapsto \omega$ at each iteration which, most of the time, cannot be computed in closed form; (ii) the algorithm iterates with respect to the ω-parameterization which is variation independent and avoids the risk of out-of-range estimates; (iii) the ω-parameterization implicitly encodes zero restrictions, in particular on β parameter for the RCG Markov property; (iv) as the Lagrange multiplier updating function does not depend on previous values, Eqs. (3.10) and (3.11) can be rephrased by a single updating function

$$\omega^{(i+1)} = \omega^{(i)} + \text{step}^{(i)} \times D^{-1}(n - \mu)(I - HPH^{\mathsf{T}}) - D^{-1}HPh.$$

At the same time, the algorithm suffers from some critical aspects, mainly related to the lack of computational efficiency when large tables are available. Firstly, the number of iterations, that rapidly increases for high dimensional tables, can be very sensitive on the starting point of the algorithm. A more sophisticated step-size method could be useful in this regard. Also, the algorithm lacks efficiency to invert matrix P at each iteration. A suitable decomposition of this matrix P could be better explored.

Note that the dimension of $\boldsymbol{P}$ depends on the size of the explanatory variables. Further algorithms for MLEs can be considered, some references with related R packages are given in the bibliographic notes of this chapter.

Remark 3.1 When the algorithm is applied for fitting a regression graph model defined by an ordering $g_1, \ldots, g_J$ of chain components, the basic factorization can be replaced by a factorization based on the parent set $\mathrm{pa}(g_j)$ of each g_j

$$p_V(\boldsymbol{\omega}) = \prod_{j=1}^{J} p_{g_j|\mathrm{pa}(g_j)}(\boldsymbol{\omega}_{g_j|g_{>j}}), \tag{3.12}$$

rather than of the predecessor set $g_{>j}$, where $\mathrm{pa}(g_j) \subseteq g_{>j}$. Then, a sequence of multivariate regression models for any $\boldsymbol{X}_{g_j}$ on $\boldsymbol{X}_{\mathrm{pa}(g_j)}$ results. Substantially, these two factorizations are equivalent in terms of inference for the $\boldsymbol{\beta}$ parameters because $\boldsymbol{X}_{g_j} \perp\!\!\!\perp \left(\boldsymbol{X}_{g_{>j}} \backslash \boldsymbol{X}_{\mathrm{pa}(g_j)}\right) | \boldsymbol{X}_{\mathrm{pa}(g_j)}$ for the RCG Markov property, they only require a different algorithm to reconstruct the joint probability vector $\boldsymbol{\pi}$ of p_V which does not represent a relevant parameter of interest in this setting (Drton 2009; Marchetti and Lupparelli 2011). When the size of $\mathrm{pa}(g_j)$ is much smaller then $g_{>j}$, considering the regression framework $\boldsymbol{X}_{g_j}$ on $\boldsymbol{X}_{\mathrm{pa}(g_j)}$ becomes more convenient to reduce the lack of efficiency to invert matrix $\boldsymbol{P}$.

3.2.2 Asymptotic Properties and Model Comparison

Under the regularity conditions assumed by Aitchison and Silvey (1958), the MLE $\mathrm{vec}(\hat{\boldsymbol{\omega}}, \hat{\boldsymbol{\tau}})$ exists with probability going to 1 as the sample size N goes to infinity. Also, the estimators of $\boldsymbol{\omega}$ and $\boldsymbol{\tau}$ are consistent, asymptotically independent and distributed as a multivariate random normal where estimate of the covariance matrix is

$$[\boldsymbol{G}^{(\infty)}]^{-1} = \begin{bmatrix} \boldsymbol{D}^{-1} - \boldsymbol{D}^{-1}\boldsymbol{H}\boldsymbol{P}\boldsymbol{H}^{\mathsf{T}}\boldsymbol{D}^{-1} & \boldsymbol{0} \\ \boldsymbol{0} & \boldsymbol{P} \end{bmatrix}.$$

As far as the uniqueness of MLEs is concerned, a general result is not available because it is difficult to establish whether a global maximum has been identified.

Aitchison and Silvey (1958) give three asymptotic equivalent goodness-of-fit statistics: the likelihood ratio (or deviance) statistic

$$G^2 = 2\boldsymbol{n}^{\mathsf{T}} \log\left(\frac{\ell(\boldsymbol{\pi}_{\mathrm{obs}})}{\ell(\boldsymbol{\pi}_{\mathrm{fit}})}\right) = 2\{\ell(\boldsymbol{\pi}_{\mathrm{obs}}; \boldsymbol{n}) - \ell(\boldsymbol{\pi}_{\mathrm{fit}}; \boldsymbol{n})\},$$

the Wald statistic

$$W^2 = \boldsymbol{h}(\boldsymbol{\omega}_{\mathrm{obs}})^{\mathsf{T}}[\mathrm{avar}(\boldsymbol{h}(\boldsymbol{\omega}_{\mathrm{obs}}))]^{-1}\boldsymbol{h}(\boldsymbol{\omega}_{\mathrm{obs}}),$$

where $\mathrm{avar}(\boldsymbol{h}(\boldsymbol{\omega}_{obs}))$ is the asymptotic variance, and the Lagrange multiplier statistic

$$L^2 = \boldsymbol{\tau}_{\mathrm{fit}}^{\mathsf{T}}[\mathrm{avar}(\boldsymbol{\tau}_{\mathrm{fit}})]^{-1}\boldsymbol{\tau}_{\mathrm{fit}}.$$

Subscripts 'obs' and 'fit' are used to denote the estimate of the parameter vector in the full and in the fitted model, respectively. Under the null hypothesis these statistics have a central chi-squared limiting distribution with degree of freedom equal to the number of parameter constraints. Penalized likelihood criteria are typically used for model comparison. For instance Akaike's information criterion (AIC) Akaike (1973)

$$\mathrm{AIC} = -2\ell(\hat{\omega}) + 2p \tag{3.13}$$

where p is the number of parameters included in a model. The model with lowest AIC should be chosen in order to select the model that provides the greatest amount of information about the observed data. The Bayesian Information Criterion (BIC) represents an AIC variant

$$\mathrm{BIC} = -2\ell(\hat{\omega}) + 2\log(N). \tag{3.14}$$

The model with lowest BIC should be chosen.

3.3   Fitting a Regression Graph Model

Assume first that a collection of blocks of variables (the chain components) is available with a selected total order of the blocks. Assume also that the data have the structure of a multidimensional contingency table, and that the sample size is sufficiently large to have no sampling zeroes in the cells.

As an example we will discuss the fitting of an adequate regression chain model to the data of the General Social Survey discussed in Example 2.10 where we have 4 categorical variables and two chain components $g_1 = \{D, G\}$, $g_2 = \{J, B\}$; see Table 3.1.

Table 3.1 Contingency table of 4 variables of the General Social Survey. Each column is labeled by $k = (j, b)$, the couples of levels of J and B

B			1			2			3		
D	G	J	1	2	3	1	2	3	1	2	3
1	1		1168	705	254	1931	1577	548	449	367	218
2			453	299	112	733	671	300	201	175	88
1	2		389	242	83	663	540	184	158	165	74
2			107	75	26	159	153	71	51	51	33

Let $\boldsymbol{\mu}$ be the vector of the expected counts of the $2 \times 2 \times 3 \times 3$ contingency table of variables X_D, X_G, X_J, X_B, in inverse lexicographic order. Let $\boldsymbol{\mu}_k$ be the subvectors of $\boldsymbol{\mu}$ with the structure

$$\boldsymbol{\mu}_k = \begin{bmatrix} \mu_{0,0,k} \\ \mu_{1,0,k} \\ \mu_{0,1,k} \\ \mu_{1,1,k} \end{bmatrix}, \tag{3.15}$$

where $k = (j, b)$, with $j, b = 1, 2, 3$, denotes a generic pair of levels of the covariates X_J and X_B. Therefore, the vectors $\boldsymbol{\mu}_k$ contain the expected joint counts of variables X_D and X_G given $(X_J = j, X_B = b)$ and they are assumed to be independently distributed as multinomials with parameters n_k and $\boldsymbol{\pi}_k$. Here n_k are the observed marginal frequencies of (X_J, X_B) and $\boldsymbol{\pi}_k$ is the vector of the conditional probabilities $P(X_D = i, X_G = j \mid (X_J, X_B) = k)$.

The link function transforms the conditional expected counts in (3.15) into 2 univariate marginal logits and 1 bivariate logit

$$\eta_{D|k} = \begin{bmatrix} 1 & -1 \end{bmatrix} \log \begin{bmatrix} \mu_{0+|k} \\ \mu_{1+|k} \end{bmatrix}, \quad \eta_{G|k} = \begin{bmatrix} 1 & -1 \end{bmatrix} \log \begin{bmatrix} \mu_{+0|k} \\ \mu_{+1|k} \end{bmatrix},$$

$$\eta_{DG|k} = \begin{bmatrix} 1 & -1 & -1 & 1 \end{bmatrix} \log \begin{bmatrix} \mu_{00|k} \\ \mu_{10|k} \\ \mu_{01|k} \\ \mu_{11|k} \end{bmatrix},$$

where we denote the sum with respect to an index with the symbol '+'. The full multivariate logistic link function is then defined by

$$\text{mlogit}(\boldsymbol{\mu}) = \begin{bmatrix} \boldsymbol{\eta}_D \\ \boldsymbol{\eta}_G \\ \boldsymbol{\eta}_{DG} \end{bmatrix} \quad \text{denoted by the formulae} \quad \begin{cases} D \sim \\ G \sim \\ D * G \sim \end{cases}$$

where $\boldsymbol{\eta}_D = [\eta_{D|k}]$, $\boldsymbol{\eta}_G = [\eta_{G|k}]$ and $\boldsymbol{\eta}_{DG} = [\eta_{DG|k}]$ are vectors of dimension $9 = |J| \times |B|$.

A regression chain graph model is defined by means of logistic regression models specifying the dependence of (X_D, X_G) on the explanatory variables (X_J, X_B),

$$\boldsymbol{\eta}_D = \boldsymbol{Z}_{BJ} \boldsymbol{\beta}_D, \quad \boldsymbol{\eta}_G = \boldsymbol{Z}_{BJ} \boldsymbol{\beta}_G, \quad \boldsymbol{\eta}_{DG} = \boldsymbol{Z}_{BJ} \boldsymbol{\beta}_{DG}, \tag{3.16}$$

where $\boldsymbol{Z}_{BJ}$ is a design matrix and $\boldsymbol{\beta}_D, \boldsymbol{\beta}_G, \boldsymbol{\beta}_{DG}$ are the regression coefficients. The design matrix is built starting from the full factorial expansion of the two explanatory variables shown in Table 3.2 with model formula

$$\sim J + B + J : B \quad \text{or equivalently with} \quad \sim J * B.$$

Table 3.2 A complete design matrix for two discrete variables X_J and X_B with 3 levels. The first column defines the constant, the columns J_2 and J_3 are the dummy variables for X_J. B_2 and B_3 are the dummies for X_B and the remaining 4 columns define the interaction term $J : B$

·	J_2	J_3	B_2	B_3	$J_2 : B_2$	$J_3 : B_2$	$J_2 : B_3$	$J_3 : B_3$
1	0	0	0	0	0	0	0	0
1	1	0	0	0	0	0	0	0
1	0	1	0	0	0	0	0	0
1	0	0	1	0	0	0	0	0
1	1	0	1	0	1	0	0	0
1	0	1	1	0	0	1	0	0
1	0	0	0	1	0	0	0	0
1	1	0	0	1	0	0	1	0
1	0	1	0	1	0	0	0	1

The saturated model is defined by (3.16) by setting Z_{BJ} equal to the complete design matrix of Table 3.2 in each regression equation, that is with the three model formulae

$$(D \sim J * B), \quad (G \sim J * B), \quad (D * G \sim J * B). \tag{3.17}$$

To fit a conditional independence model $X_D \perp\!\!\!\perp X_B \mid X_J$ we need to remove the arrow $D \leftarrow B$. The correct model formulae are

$$(D \sim J), \quad (G \sim J * B), \quad (D * G \sim J * B) \tag{3.18}$$

obtained by simplifying the first model formula, and leaving unchanged the remaining ones. Another interesting reduced model is obtained by removing the dependence of the bivariate logit η_{DG} on the explanatory variables. In this case the design matrix Z_{BJ} of the third regression equation in (3.16) reduces to a column of ones with a model formula

$$(D \sim J * B), \quad (G \sim J * B), \quad (D * G \sim 1). \tag{3.19}$$

A further simplification is to delete the interaction terms $J * B$ by assuming that the logistic regressions are additive:

$$(D \sim J + B), \quad (G \sim J + B), \quad (D * G \sim 1) \tag{3.20}$$

Notice that the last two models in Eqs. (3.19) and (3.20) introduce non-independence constraints with the intent to simplify the model and to improve, at the same time, the model fitting as well as the parameter interpretation.

Example 3.1 (Analysis of GSS data of Table 3.1) Given the list of chain components and their ordering, the process of fitting is reduced to a strategy to test research

Table 3.3 Regression chain graph models fitted

	Model	G^2	df	p
(1)	$D * G \sim 1$	4.926	8	0.765
(2)	$(1) + (D \sim J + B), (G \sim J + B)$	16.067	16	0.448
(3)	$(1) + (D \sim J), (G \sim B)$	19.240	20	0.506
(4)	$(1) + (D \sim J), (G \sim 1)$	26.490	22	0.231

Table 3.4 Estimates of the parameters of the models (2) and (3)

Model (2)	D			G			$D * G$		
	β	se	p	β	se	p	β	se	p
Const.	-1.040	–	–	-1.190	–	–	-0.371	0.05	0.000
J_2	0.099	0.04	0.013	0.041	0.04	0.305			
J_3	0.251	0.06	0.000	-0.010	0.06	0.868			
B_2	0.004	0.04	0.920	-0.005	0.05	0.920			
B_3	0.080	0.06	0.182	0.141	0.06	0.019			
Model (3)	β	se	p	β	se	p	β	se	p
Const.	-1.028	–	–	-1.178	–	–	-0.371	0.05	0.000
J_2	0.103	0.04	0.01						
J_3	0.256	0.06	0.00						
B_2				-0.003	0.05	0.952			
B_3				0.147	0.06	0.014			

hypotheses concerning the problem under study. These hypotheses are necessarily constrained by the chosen order of chain components and the structure of the regression graph. First we can try to reduce the number of parameters of the saturated model in search of a simple starting model.

The first step is usually that of testing the presence of the interaction terms in the 3 equations, that is the models (3.19) and (3.20). Table 3.3 reports for these two fitted model the likelihood ratio test G^2 corresponding to the difference between the deviance of the model (that is, minus twice the maximized log-likelihood of the model) and the deviance of the saturated model, the degrees of freedom (d.f.) and the p-value (p). The results show that the model (1) stating that the association between the opinions D, G is constant is not rejected, according to the asymptotic chi-squared distribution of the likelihood ratio statistic $G^2 = 4.926$ on 8 degrees of freedom. From the estimates of the parameters of model (2) in Table 3.4 appears that several terms are not significant; specifically, in the model for the opinion D the effect of confidence in bank B and the effect of job satisfaction J on the opinion G.

Thus a further simplification can be attempted by removing the parameters associated to the edge $D \leftarrow B$ or to the edge $G \leftarrow J$ or both. Choosing the latter case, we fit the model (3) consisting of model (1) plus $(D \sim J), (G \sim B)$ obtaining a $G^2 = 19.234$ with 20 d.f., see Table 3.3. Thus the likelihood ratio test statistic for com-

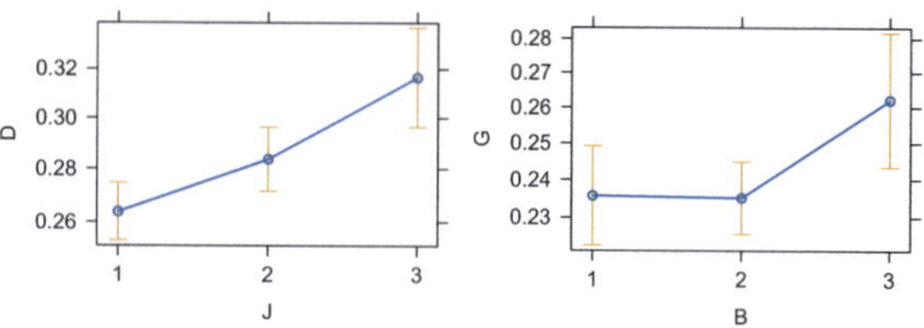

Fig. 3.1 Plots to evaluate the change of probabilities of being against the death penalty for murder (to the left) and against the permit for guns (to the right)

paring model (3) with model (2) is $w = 19.240 - 16.067 = 3.173$ with $20 - 16 = 4$ degrees of freedom that is clearly not significant.

By looking at the estimates of Model (3) on could be tempted to remove also the edge $G \leftarrow B$ thus obtaining a model (4) with test statistic G^2 reported in Table 3.3. However, by comparison with model (3) the likelihood ratio test statistic $w = 26.49 - 19.24 = 7.25$ with 2 degrees of freedom is significant (p-value $= 0.027$) and causes the model to be rejected.

The best model selected by this procedure is therefore model (3). The first interpretation is that the opinions on death penalty for murder, D, and on requiring a permit before buying a gun, G, are negatively associated with a fitted conditional log odds-ratio -0.371 not depending on the job satisfaction, J, and confidence in banks, B. Second, there is a dependence of opinion D on death penalty just on J given B and a dependence of opinion on permit for guns G only on B given J. By considering the ordinal nature of the opinions on job satisfaction it can be seen from the estimates of the logistic regression for the response D that when job satisfaction decreases the opinion against death penalty increases (remember that D has levels $1 = $ in favor, $2 = $ against). For the response G, when confidence in banks decreases the opinion against the permit for buying guns increases. Figure 3.1 shows the two effect plots. In the end, the final model chosen has the graph representation shown in Fig. 2.2. The regression graph interpretation is limited to the two conditional independencies $X_D \perp\!\!\!\perp X_B \mid X_J$ and $X_G \perp\!\!\!\perp X_J \mid X_B$, but in fact the final model contains a non-independence constraint concerning the removal of the dependence of $D * G$ on J, B. A model defining only the above two conditional independencies would be instead

$$(D \sim J), \quad (G \sim B), \quad (D * G \sim J * B).$$

Before concluding the analysis, we could also be interested in testing the hypothesis that the two responses are independent given the covariates, i.e., $X_D \perp\!\!\!\perp X_G \mid (X_J, X_B)$. This independence can be expressed by $\eta_{DG} = \mathbf{0}$ a constraint that can be inserted directly into Eq. (3.16). We will denote symbolically this constraint as $D * G \sim 0$. Therefore, the pure conditional independence model $X_D \perp\!\!\!\perp X_G \mid (X_J, X_B)$ is written as

$$(D \sim J * B), \quad (G \sim J * B), \quad (D * G \sim 0).$$

This model is however rejected against the saturated model, given a quite high $G^2 =$ 68.37, with 9 degrees of freedom. In any case the well-fitting regression graph model corresponds to that in Fig. 2.2. $\qquad\square$

3.4 Strategies for Structure Learning of Regression Chain Graphs

In the previous example the four variables were grouped in two chain components with a clear ordering. The first block groups opinions on civil liberties, and the second contains the predictors job satisfaction and confidence in banks. Things get more complicated when there are several variables and poor prior knowledge because the variables can be partitioned in different ways and with different orderings.

Notice that in general, it seems advisable to include in the chain components a limited number of variables. This helps to limit the difficulty of estimating models with many response variables and also improves model selection.

Example 3.2 (Choice of the chain components) Consider two further variables belonging to the GSS dataset, that is $A =$ the opinion about the possibility of abortion if pregnant as a result of rape (1 = favor, 2 = oppose), and $S =$ the gender of the respondent (1 = male, 2 = female). While gender is a surely a background variable, the opinion on abortion is a variable rather different from the other two responses that can also be treated as a separate intermediate variable. For instance, we could consider the following partitions:

$$
\begin{array}{ll}
(1) \quad \{A, D, G\}, \{B\}, \{J, S\} & (2) \quad \{D, G\}, \{A\}, \{B, J\}, \{S\} \\
(3) \quad \{A\}, \{D, G\}, \{B\}, \{J, S\} & (4) \quad \{A, G\}, \{D\}, \{B, J\}, \{S\}
\end{array}
$$

where the ordering of the blocks is from left to right. Each option has a different justification possibly dictated by a specific research hypothesis. For instance partition (1) has a block with 3 joint primary responses concerning opinions on civil rights, (2) considers $\{D, G\}$ as primary responses and $\{A\}$ as a secondary response, and both the economic-financial variables as intermediate variables. Finally, both (3) and (4) choose to have a first block containing the variable A. $\qquad\square$

In principle, with many variables the number of possible partitions is likely to become quite large. But the problem is mitigated if we are able to specify the order between the blocks of variables that are considered on equal standing, while simultaneously trying to find empirically an ordering of such blocks satisfying the greatest number of conditional independencies.

One possible approach is that of searching for conditional independencies represented as a DAG where the vertices are the chain components. To do this, we define for each block g_j, $j = 1, \ldots, J$, a single categorical variable C_j obtained by combining the levels of the variables in the block with the associated joint probabilities.

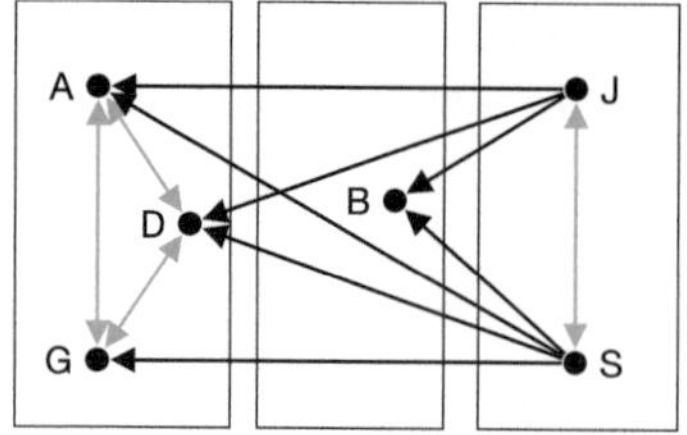

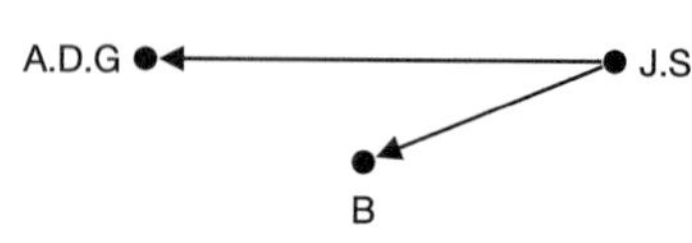

Fig. 3.2 Left: a regression chain graph. Right: the implied DAG of chain components

Example 3.3 (Combined variables) The chain component $\{J, B\}$ where both B and J have levels 1, 2, 3, yields a combined variable $J \cdot B$ with levels 1.1, 2.1, 3.1, 1.2, 2.2, 3.2, 1.3, 2.3, 3.3. With the chain component $\{D, G\}$ the combined variable $D.G$ has levels 1.1, 2.1, 1.2, 2.2. The observed frequencies of the joint distribution of $D.G$ and $J.B$ are contained in Table 3.1, reshaped as a 4×9 table instead of a $2 \times 2 \times 3 \times 3$ array. $\qquad\square$

Then, we define a DAG having as vertices the combined variables C_j. Every pair of vertices C_i, C_j is connected by an arrow by respecting the ordering of the chain components g_i and g_j, except when C_i and C_j are independent given all the preceding combined variables. The resulting graph is called the *DAG of chain components*.

Example 3.4 (The DAG of chain components) Assume that the graph to the left of Fig. 3.2 is the true regression graph. Define for the first and third chain components two variables $A.D.G$, and $J.S$ where the operator '.' indicates the combination of the levels of the variables. The true chain graph implies a conditional independence $(X_A, X_D, X_G) \perp\!\!\!\perp X_B \mid (X_J, X_S)$ that is equivalent to $X_{A.D.G} \perp\!\!\!\perp X_B \mid X_{J.S}$ and thus the DAG of the chain components is drawn as in Fig. 3.2 to the right. Notice that the implied DAG would be the same if the bi-directed graphs within the blocks were incomplete or if some arrows (but by no means all) were missing. $\qquad\square$

Of course we don't know the structure of the true regression graph but, given the knowledge of the chain components and the data, it is possible to operate a search in the space the DAGs of chain components trying to *learn the DAG structure*.

Learning DAGs is a highly sophisticated automatic procedure of model selection that it is not based just on classic "backward" or "forward" regression, but on complex and efficient algorithms that select the best graphs on the basis of appropriate criteria, by searching in the very large class of all possible DAGs.[1] Simplifying, there are two main DAG structure learning algorithms: (a) constraint-based, and (b) score-based; see Scutari and Denis (2021).

(a) *Constraint-based algorithms* are based on performing several conditional independence tests with the goal of finding the DAGs that are compatible with

---

[1] Notice however that the class of all DAGs is superexponential as a function of the number of vertices.

those independencies. The most widely known algorithm in this class is the PC-algorithm with its variants.

(b) *Score-based algorithms* instead consider each DAG as a whole and assign a score to each candidate network. Then they try to maximize the score using a heuristic search, such as hill-climbing. The most used network scores are the Akaike or the Bayesian Information Criterion defined in Eqs. (3.13) and (3.14).

Both strategies are made available in R with the packages **pcalg** (Markus Kalisch et al. 2012) for (a) and **bnlearn** (Scutari 2010) for (b). The latter is especially recommended for the ease of use with a wide choice of algorithms, tests and scores.

The conditional independence test mentioned for constraint-based algorithms can be obtained with the log-likelihood ratio statistic G^2 or other variants. Given the hypothesis $X_A \perp\!\!\!\perp X_B \mid X_C$, the G^2 is defined by

$$
G^2 = \sum_{a=1}^{|A|} \sum_{b=1}^{|B|} \sum_{c=1}^{|C|} \frac{n_{abc}}{n_{+++}} \log \frac{n_{abc} n_{++c}}{n_{a+c} n_{+bc}},
\tag{3.21}
$$

where n_{abc} is the observed count in a generic cell (a, b, c) of the three-way table $A \times B \times C$ and the symbol $|\cdot|$ denotes here the number of levels of the categorical variables. The G^2 test has an asymptotic chi-squared distribution with degrees of freedom d.f. $= (|A| - 1)(|B| - 1)|C|$ under the null hypothesis of conditional independence. Marginal independence tests such as $X_A \perp\!\!\!\perp X_B$ can also be done by a suitably modified version of G^2, by setting $|C| = 1$.

In using these algorithms is essential to include any available prior information concerning the groups of variables to be treated jointly and the partial order of the variables depending on their role (background, intermediate, primary responses). For instance, this can be done by imposing *black lists* containing the arrows that are forbidden.

Example 3.5 (Searching for conditional independencies) We shall apply a hill-climbing algorithm based on the AIC score to the combined variables of the 4 partitions of the Example 3.2. During the search we systematically used a black list for the arrows that do not respect the assumed order. The resulting DAGs are shown in Fig. 3.3. We remark that this procedure based on searching a DAG for the sets of combined variables does not have the ambition of proposing an ordering for single variables, which is a well-known issue in the field of DAG selection. Our intent is rather to leverage from the comparison of models selected under different orderings, a priori defined, that address different research hypotheses.

The associated conditional independencies, with the test statistics G^2 with the appropriate degrees of freedom are listed below[2]:

---

[2] Recall that in chi-squared distributions values smaller than the degrees of freedom occur with probability ≥ 0.5, indicating a satisfactory fit under the given assumptions.

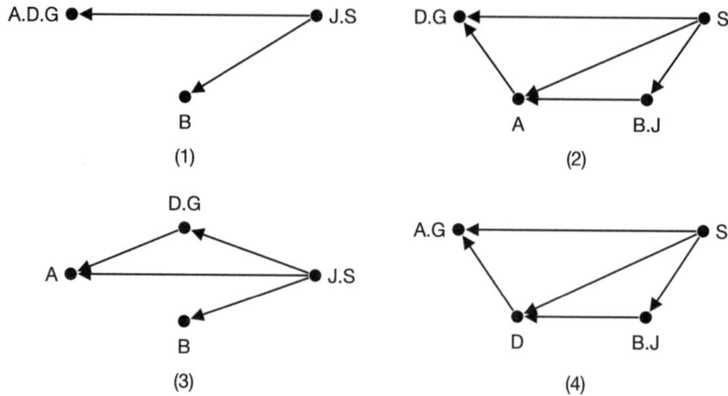

Fig. 3.3 The 4 DAGs of chain components associated with the blocks of variables of Example 3.2, obtained from the hill-climbing structure learning algorithm with AIC score

$$(1) \quad (X_A, X_D, X_G) \perp\!\!\!\perp X_B \mid (X_J, X_S), \qquad\qquad G^2_{84} = 77.9$$

$$(2) \quad (X_D, X_G) \perp\!\!\!\perp (X_B, X_J) \mid (X_A, X_S), \qquad\qquad G^2_{96} = 108.6$$

$$(3) \quad (X_D, X_G) \perp\!\!\!\perp X_B \mid (X_J, X_S), \qquad\qquad\quad G^2_{36} = 32.4$$

$$\qquad\quad X_A \perp\!\!\!\perp X_B \mid (X_D, X_G, X_J, X_S), \qquad\qquad G^2_{48} = 45.5$$

$$(4) \quad (X_A, X_G) \perp\!\!\!\perp (X_B, X_J) \mid (X_D, X_S), \qquad\qquad G^2_{96} = 110.5.$$

All the test statistics are non-significant leaving some freedom in the choice of components and their order. It can be verified that the conditional independence tests of the DAGs (1) and (3) are equivalent and the global goodness of fit statistics are the same. For the DAG (3) the full G^2 turns out to be the sum $G^2_{36} + G^2_{48} = 32.4 + 45.5 = 77.9 = G^2_{84}$.

An analogous search has been also explored for the regression graph model in Fig. 2.4 where the ordered components are $\{D\}, \{A, B, J, G, \}, \{S\}$. This example is not treated in depth since the search provides a full DAG model for the chain components. $\qquad\qquad\qquad\qquad\qquad\qquad\qquad\qquad\qquad\qquad\qquad\qquad\qquad\quad \square$

The DAG of chain components is a rough simplification of the structure of a regression graph, which takes for granted the joint dependence of all variables within the blocks. The essential further step is the selection of reduced regression models for each conditional distribution $p_{g_j \mid g_{>j}}$ that can be carried out as suggested in Sect. 3.3.

Example 3.6 (Results of model selection for the GSS data) We give all the details of the regression chain graph model obtained starting from the DAG of chain components (1). Two multivariate logistic regression models with responses (D, A, G), (B) are fitted and then reduced by removing the non-significant predictors. Further, it is verified that the variables J and S in the last block are independent.

Figure 3.4 (left) shows the final well-fitting model, where the starred regression equations correspond to conditional independence restrictions that imply missing

D, A, G	B	J, S
$D \sim J + S$	$B_2 \sim J + S$	$J_2 \sim 1$
$A \sim J + S$	$B_3 \sim J + S$	$J_3 \sim 1$
$\star\, G \sim S$		$S \sim 1$
$D * A \sim 1$		$\star\, J_2 : S \sim 0$
$D * G \sim 1$		$\star\, J_3 : S \sim 0$
$\star\, A * G \sim 0$		
$D * A * G \sim 0$		

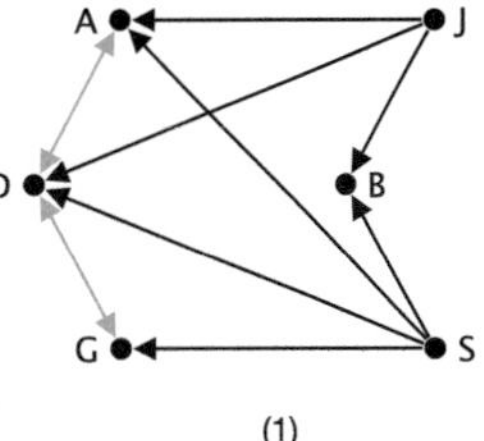

(1)

Fig. 3.4 Left: the 3 sets of reduced logistic models in symbolic form for the regression graph (1). The 3 starred models are equivalent to 3 conditional independencies $X_G \perp\!\!\!\perp X_J \mid (X_B, X_S)$, $X_A \perp\!\!\!\perp X_G \mid (X_B, X_J, X_S)$ and $X_J \perp\!\!\!\perp X_S$. Right: the final regression graph (1)

Table 3.5 Likelihood ratio test statistic G^2 for the regression chain graph (1). The rows correspond to the DAG of the chain components, the pure regression graph defined by conditional independencies only, and by the further non-independence constraints of Fig. 3.4 (left)

Model (1)	G^2	df
DAG of chain components	77.9	84
Pure regression graph	91.6	96
Final regression graph	114.8	120

edges. The remaining equations contain non-independence restrictions that is constraints that simplify the dependence structure, by removing non-significant interactions. The associated regression graph is shown in Fig. 3.4.

Table 3.5 shows the goodness of fit statistic G^2 for a sequence of three nested models: the starting DAG of the chain components (1), the pure regression graph found with only conditional independence restrictions and the final reduced model with additive predictors. The maximum likelihood estimates of the parameters are shown in Table 3.6.

From the regression parameters we have the opportunity to discover the interpretation of the dependencies. The opinions concerning death penalty and abortions depend additively on job satisfaction and gender. The opposition to death penalty tends to increase for women and the lower the work satisfaction. Men more than women tend to be in favor of abortion as a result of rape. The opposition to abortion is larger when the job satisfaction is high. Women are more supportive of gun regulation than men.

The technique used above can also be applied to the remaining DAGs of chain components (2)–(4) in Fig. 3.3. Leaving out the details, we will just show the final regression graphs, the final models with the goodness of fit (omitting the pure regression graph), and eventually we make some comparisons. Table 3.7 contains the reduced models and the goodness of fit and Fig. 3.5 shows the associated regression graphs.

The regression graphs (1) and (3) are rather similar but contain important differences. (1) has less edges and looks simpler but has 3 joint responses, while (3) has more edges but has a single main response A (opinion about abortion) and 2 joint

Table 3.6 Estimates of the parameters of model in Fig. 3.4

D	β	se	A	β	se	G	β	se
Const.	−1.3067	0.0381	Const.	−1.5813	0.0429	Const.	−0.7658	0.0284
J_2	0.1061	0.0430	J_2	−0.2118	0.0513	S_2	−0.7664	0.0417
J_3	0.2562	0.0571	J_3	−0.2443	0.0714			
S_2	0.4574	0.0403	S_2	0.1718	0.0475			
$D * A$	β	se	$D * G$	β	se			
Const.	0.5585	0.0494	Const.	−0.3012	0.0489			
B_2	β	se	B_3	β	se	$J * S$	β	se
Const.	0.5041	0.0364	Const.	−0.7119	0.0502			
J_2	0.2930	0.0437	J_2	0.3393	0.0619	J_3	−0.2519	0.0191
J_3	0.3388	0.0623	J_3	0.7551	0.0798	J_3	−1.1764	0.0260
S_2	−0.0035	0.0406	S_2	−0.3482	0.0560	S_2	0.2475	0.0176

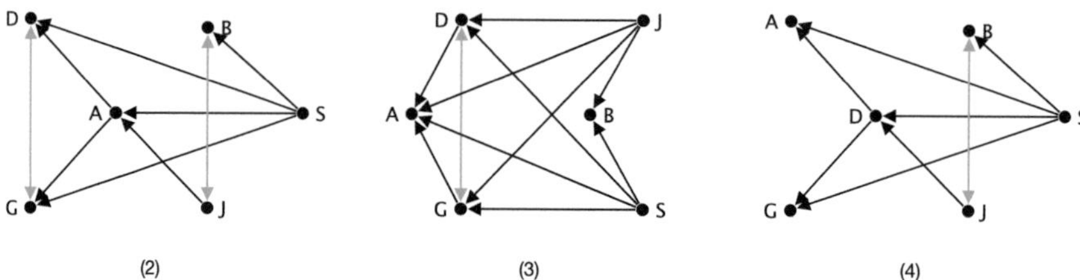

Fig. 3.5 The final regression chain graphs deduced from the DAGs of chain components (2)–(4) of Fig. 3.3

intermediate variables D, G (opinions about death penalty and guns) in the second block. The model fits are $G^2_{120} = 114.8$ versus $G^2_{116} = 104.2$ and thus for both models the observed chi-squared statistics are smaller than the degrees of freedom. A relevant feature is that the background variables X_J (job satisfaction) and X_S (gender) are marginally independent.

On the other hand, the regression graphs (2) and (4) are also similar but both have chi-squared statistic larger than the degrees of freedom, although not significant: $G^2_{122} = 137.1$ and $G^2_{121} = 138.8$. An interesting aspect is that the joint variables X_B and X_J (both economic variables) in both graphs are dependent but confidence in banks is weakly associated with gender only through the third class B_3. □

As a conclusion, the suggested strategy combines an automatic procedure of model selection (but guided by a given order of the variables), with a reasoned approach of reduction of regression models based on substantive research hypotheses. This strategy is designed to avoid automatic procedure that necessarily lead to a single model, when in fact, as in our case, there are models that fit equally well.

Table 3.7 The logistic regression models relative to the DAGs of chain components (2)–(4), with the log-likelihood ratio statistics

Model (2)

DAG of chain components: $G^2_{96} = 108.6$. Final regression graph: $G^2_{122} = 137.1$

D, G	A	B, J
$D \sim A + S$	$A \sim J + S$	$B_2 \sim 1, B_3 \sim S, J_2 \sim 1, J_3 \sim 1$
$G \sim A + S$		$B_2 : J_2 \sim 1, B_3 : J_2 \sim 1, B_2 : J_3 \sim 1, B_3 : J_3 \sim 1$
$D * G \sim 1$		

Model (3)

DAG of chain components: $G^2_{84} = 77.9$. Final regression graph: $G^2_{116} = 104.2$

A	D, G	B	J, S
$A \sim D + G + J + S$	$D \sim J + S$	$B_2 \sim J$	$J_2 \sim 1, J_3 \sim 1, S \sim 1$
	$G \sim J + S$	$B_3 \sim J + S$	$J_2 : S \sim 0, J_3 : S \sim 0$
	$D * G \sim 1$		

Model (4)

DAG of chain components: $G^2_{96} = 110.5$, Final regression graph: $G^2_{121} = 138.8$

A, G	D	B, J
$A \sim D + S$	$D \sim B + J + S$	$B_2 \sim 1, B_3 \sim S, J_2 \sim 1, J_3 \sim 1$
$G \sim D + S$		$B_2 : J_2 \sim 1, B_3 : J_2 \sim 1, B_2 : J_3 \sim 1$
$A * G \sim 0$		$B_3 : J_3 \sim 1$

3.5 Appendix: Hessian Matrix Approximations

A simplified Newton-Raphson algorithm has been proposed by Aitchison and Silvey (1958); this version does not require to compute matrix $\boldsymbol{F}^{(i)}$ at each i-iteration, but only a single inversion of matrix $\boldsymbol{F}^{(0)}$

$$\boldsymbol{F}^{(0)} = \begin{bmatrix} -\boldsymbol{D}^{(0)} & \boldsymbol{H}^{(0)} \\ \cdot & \boldsymbol{0} \end{bmatrix}$$

computed assuming a reasonable starting estimates $\boldsymbol{\omega}^{(0)} = \log(\boldsymbol{n})$ and $\boldsymbol{\tau}^{(0)} = 0$. The adjustment $\boldsymbol{\omega}^{(0)} = \log(\boldsymbol{n}) + \boldsymbol{\varepsilon}$ with a small $\boldsymbol{\varepsilon}$ is typically used for the sampled zero cell counts.

Lang (1996) also suggests a modified Newton-Raphson iterative scheme with an approximation of the Hessian matrix easier to be inverted in each iteration. It is proved that, as $\boldsymbol{\tau}$ is close to $\hat{\boldsymbol{\tau}}$ and as N gets larger,

$$\frac{d\boldsymbol{H}}{d\boldsymbol{\omega}^T}(N^{-1}\boldsymbol{\tau} \otimes \boldsymbol{I}_s) = o(1)$$

where s is the number of constraints and

$$N^{-1}\boldsymbol{F} = N^{-1}\boldsymbol{G} + \begin{bmatrix} o(1) & 0 \\ 0 & 0 \end{bmatrix},$$

with $o(1)$ denoting an asymptotically zero value. Then, the Hessian matrix $\boldsymbol{F}^{(i)}$ at i-iteration can be approximated with matrix $\boldsymbol{G}^{(i)}$

$$\boldsymbol{G}^{(i)} = \begin{bmatrix} -\boldsymbol{D}^{(i)} & \boldsymbol{H}^{(i)} \\ \cdot & \boldsymbol{0} \end{bmatrix}$$

where its inverse is

$$(\boldsymbol{G}^{(i)})^{-1} = \begin{bmatrix} \boldsymbol{D}^{-1} - \boldsymbol{D}^{-1}\mathbf{H}\mathbf{P}\mathbf{H}^{\mathsf{T}}\boldsymbol{D}^{-1} & \boldsymbol{D}^{-1}\mathbf{H}\mathbf{P} \\ \cdot & \boldsymbol{P} \end{bmatrix}.$$

where $\boldsymbol{P} = (\boldsymbol{H}^{\mathsf{T}}\boldsymbol{D}^{-1}\boldsymbol{H})^{-1}$ is a symmetric positive definite matrix.

3.6 Bibliographic Notes

Section 3.1

Several algorithms are available for maximum likelihood inference of categorical data under equality constraints. Some of them are based on the maximization of the Lagrangian log-likelihood with an iterative scheme inspired by Aitchison and Silvey (1958); see Lang (1996) for technical details details on the maximization procedure, the asymptotic properties of MLEs and Colombi and Forcina (2001), Evans and Forcina (2013) who complement the algorithm with the use of Fisher scoring method. The Iterative Proportional Fitting algorithm represents another well-established approach for MLEs of contingency table models and it can be also applied for fitting regression graph models; see Rudas (1998) and Rudas et al. (2010).

Section 3.2

The algorithm proposed in this chapter is inspired to Bergsma (1997) who also developed a method to iteratively define the step during the maximization scheme. Drton and Richardson (2008) proposed an approach based on the Iterative Conditional Fitting for binary bi-directed graph models that, in principle, can be also generalized to regression graph models. Asymptotic properties for MLEs of discrete bi-directed graph and regression graph models are discussed in Lupparelli (2006), Lupparelli et al. (2009) and Marchetti and Lupparelli (2011).

Section 3.3

There are some R packages for MLEs of categorical data. For the fitting procedure described in this chapter we specifically refer to the package **cta** (Lang 2005). The package **cmm** (Bergsma et al. 2009) can be also used for fitting marginal models for categorical data, regardless of the context of regression graph models. Hierarchical multinomial models under equality and inequality constraints for both categorical and ordinal data (Bartolucci et al. 2007) can be fitted by using the package **hmmm** (Colombi et al. 2014). The plot in Fig. 3.1 is done using the R package **effects**; see Fox and Weisberg (2018, 2019).

Section 3.4

Model selection for regression graph models has been barely explored in the literature. Sonntag and Peña (2012) and Javidian and Valtorta (2021) proposed methods for learning the structure of regression chain graphs, to which the interested reader can refer. In this chapter a model search procedure is proposed based on the selection of the DAG of the chain component induced by the basic factorization; the DAG search has been performed by using the R package **bnlearn** (Scutari and Denis 2021).

Chapter 4
Bayesian Inference

Summary

Bayesian inference for discrete regression chain graph models is discussed, specifically for graphs that are Markov equivalent to bi-directed graphs, that is they define the same set of independencies for the joint probability distribution as illustrated in Chap. 2. Conjugate analysis is applicable only to specific graphical configurations, requiring Markov Chain Monte Carlo techniques for broader applications. The discussion will encompass model specification and estimation, focusing on probability and marginal log-linear parameterizations of the model. The previous issues will be illustrated through real data applications. We additionally discuss the complexities involved in prior specification that is an essential aspect of the Bayesian framework.

4.1 Basic Background

4.1.1 Parametric Inference

In Bayesian inference, unknown parameters are treated as random variables, each associated with a probability distribution known as the *prior distribution* $f(\boldsymbol{\theta})$, which represents the initial beliefs about the parameter vector before observing any data and models its uncertainty in the parameter space Θ.

The central objective of Bayesian inference is to learn from empirical observations by determining the *posterior distribution* $f(\boldsymbol{\theta}|\boldsymbol{y})$, which is the probability distribution of the parameters $\boldsymbol{\theta}$ given the observed data $\boldsymbol{y}$. This learning mechanism follows directly from *Bayes theorem*, expressed as

$$f(\boldsymbol{\theta}|\boldsymbol{y}) = \frac{f(\boldsymbol{y}|\boldsymbol{\theta})f(\boldsymbol{\theta})}{f(\boldsymbol{y})} \propto f(\boldsymbol{y}|\boldsymbol{\theta})f(\boldsymbol{\theta}), \quad \boldsymbol{\theta} \in \Theta. \tag{4.1}$$

M. Lupparelli et al., *Regression Graph Models for Categorical Data*,
SpringerBriefs in Statistics, https://doi.org/10.1007/978-3-031-99797-6_4

The posterior distribution, $f(\boldsymbol{\theta}|y)$ is then obtained by updating the prior knowledge based on the information provided by the data through the likelihood function $f(y|\boldsymbol{\theta})$. Function $f(y)$ is called the *marginal likelihood* and is obtained by integrating over the entire parameter space the numerator of (4.1)

$$f(y) = \int_{\Theta} f(y|\boldsymbol{\theta}) f(\boldsymbol{\theta}) d\boldsymbol{\theta}, \quad \boldsymbol{\theta} \in \Theta. \tag{4.2}$$

The specification of the prior distribution is a critical aspect in Bayesian inference. When background information is available, it should be incorporated into the prior distribution. Otherwise, in the absence of prior knowledge, the prior should be chosen so that the data primarily influence the behavior of the posterior distribution.

When possible, *conjugate prior* distributions provide a convenient solution, as they preserve the same distributional family between the prior and posterior, facilitating closed-form posterior updates and simplifying computations. In many cases, there is a one-to-one correspondence between the probability distribution of the data and the conjugate priors for its parameters. For instance, when data follow a Gaussian distribution with known variance, the natural conjugate prior for the mean parameter μ is also a Gaussian distribution. The prior mean and variance, which encode prior beliefs about μ, are referred to as hyperparameters. Similarly, for a Bernoulli model, the conjugate prior for the success probability π is a Beta distribution, characterized by two shape parameters that serve as hyperparameters. In both cases, the posterior distributions of μ and π remain Gaussian and Beta, respectively, with their hyperparameters updated by using suitable statistics computed on the observed data.

Conjugate priors, while mathematically convenient, may not always be appropriate or feasible, particularly in complex models where greater flexibility in the prior distribution is required. In some cases, the probability data distribution itself may not correspond to any conjugate prior, imposing alternative approaches. When an analytical expression for the posterior is not available, Bayesian inference relies on simulation-based methods such as Markov Chain Monte Carlo (MCMC) or variational inference. These techniques enable inference in high-dimensional and computationally challenging settings where exact solutions are infeasible.

4.1.2 Model Comparison

In Bayesian inference, the model space is also treated as stochastic, and prior distributions $f(\mathcal{G}_k)$ can be specified over a set of competing models $\mathbb{G} = \{\mathcal{G}_k\}_{k \in K}$ for the observed data y. The fundamental idea is to select the model with the highest posterior probability $f(\mathcal{G}_k|y)$, which is computed using *Bayes theorem*:

$$f(\mathcal{G}_k|y) = \frac{f(y|\mathcal{G}_k) f(\mathcal{G}_k)}{f(y)} \propto f(y|\mathcal{G}_k) f(\mathcal{G}_k), \quad \mathcal{G}_k \in \mathbb{G}, \tag{4.3}$$

where $f(\boldsymbol{y}|\mathcal{G}_k)$ is the marginal likelihood under model $\mathcal{G}_k$, and $f(\boldsymbol{y})$ is the normalizing constant obtained by integrating the numerator over all models in $\mathbb{G}$. In the absence of prior information, a uniform prior over $\mathbb{G}$ is typically assumed, in which case the posterior probability in (4.3) depends only on the marginal likelihood $f(\boldsymbol{y}|\mathcal{G}_k)$.

One of the key instruments for model comparison is the *Bayes Factor* (BF), which is defined as

$$\mathrm{BF}_{kk'}(\boldsymbol{y}) = \frac{f(\boldsymbol{y}|\mathcal{G}_k)}{f(\boldsymbol{y}|\mathcal{G}_{k'})}, \qquad \mathcal{G}_k, \mathcal{G}_{k'} \in \mathbb{G}. \tag{4.4}$$

The Bayes factor quantifies the relative evidence provided by the data in favor of model $\mathcal{G}_k$ over model $\mathcal{G}_{k'}$. When both models are assigned equal prior probabilities, the BF corresponds to the ratio of their posterior probabilities. Unlike likelihood-ratio tests, the Bayes factor does not depend on specific parameter estimates; instead, it integrates over all possible parameter values within each model, properly accounting for prior uncertainty. An important property of the Bayes factor is that it satisfies the *coherence condition*: given two models $\mathcal{G}_k, \mathcal{G}_{k'} \in \mathbb{G}$,

$$\mathrm{BF}_{kk'}(\boldsymbol{y}) = \mathrm{BF}_{kk''}(\boldsymbol{y}) \times \mathrm{BF}_{k''k'}(\boldsymbol{y}), \qquad k, k', k'' \in K, \tag{4.5}$$

so that the comparison between model $\mathcal{G}_k$ and $\mathcal{G}_{k'}$ is invariant across the model space for any $\mathcal{G}_{k''} \in \mathbb{G}$. Moreover, the posterior probability in (4.3) can be expressed in terms of Bayes factor as follows:

$$f(\mathcal{G}_k|\boldsymbol{y}) = \left(1 + \sum_{k' \neq k} \frac{f(\mathcal{G}_{k'})}{f(\mathcal{G}_k)} \mathrm{BF}_{k'k}(\boldsymbol{y})\right)^{-1}, \qquad \mathcal{G}_k, \mathcal{G}_{k'} \in \mathbb{G}. \tag{4.6}$$

This formulation highlights how BFs govern the relative weighting of competing models in Bayesian model selection.

When dealing with model comparison, using *compatible priors* becomes crucial to ensures consistency across models by assigning the same prior distribution to shared marginal structures; the following example is illustrative of the basic principles of compatible priors.

Example 4.1 (Construction of Compatible Dirichlet Priors) Let us consider four binary variables A, B, C, D with joint probability distribution

$$\pi_{abcd} = P(A = a, B = b, C = c, D = d),$$

with a, b, c, d taking values in $\{0, 1\}^4$. A conjugate prior on $\boldsymbol{\pi} = (\pi_{0000}, \ldots, \pi_{1111})^{\mathsf{T}}$ is a *Dirichlet prior* $\boldsymbol{\pi} \sim \mathrm{Dirichlet}(\boldsymbol{\alpha})$, where $\boldsymbol{\alpha} = (\alpha_{0000}, \ldots, \alpha_{1111})^{\mathsf{T}}$ is the vector of hyperparameters. To ensure meaningful model comparison, prior distributions must be compatible across models. Suppose to consider the following two marginal independence models for the joint probability distribution:

(i) Model 1: $p_{ABCD} = p_{AB} \times p_{CD}$,
(ii) Model 2: $p_{ABCD} = p_{ABC} \times p_{D}$,

where π_{AB++}, π_{++CD}, π_{ABC++} and π_{+++D} represent the probability vectors of marginal distributions involved in factorizations of Models 1 and 2.

Compatibility among prior distributions is required if we are interested in comparing Model 1 and 2. Specifically, the prior distribution assigned to π_{AB++} in Model 1 must be equivalent to the prior distribution assigned to π_{AB+} in Model 2, after integrating out C from π_{ABC+}. Similarly, the prior specified on π_{++CD} in Model 1, when integrated over C, should provide the same information as the prior on π_{+++D} in Model 2. Such compatibility can be ensured by deriving these prior distributions via marginalization from a global prior, assuming a conjugate Dirichlet prior on the full model, $\pi \sim \text{Dirichlet}(\alpha)$. We show below that the prior distributions in marginal models are coherently derived. Specifically, in Model 1 we have

$$\pi_{AB++} \sim \text{Dirichlet}(\alpha_{AB++}) \quad \text{with} \quad \alpha_{ab++} = \sum_{c,d} \alpha_{abcd},$$

$$\pi_{++CD} \sim \text{Dirichlet}(\alpha_{++CD}) \quad \text{with} \quad \alpha_{++cd} = \sum_{a,b} \alpha_{abcd},$$

and in Model 2

$$\pi_{ABC+} \sim \text{Dirichlet}(\alpha_{ABC+}) \quad \text{with} \quad \alpha_{abc+} = \sum_{d} \alpha_{abcd},$$

$$\pi_{+++D} \sim \text{Dirichlet}(\alpha_{+++D}) \quad \text{with} \quad \alpha_{+++d} = \sum_{a,b,c} \alpha_{abcd}.$$

$\square$

Example 4.1 shows that the use of compatible priors prevents artificial biases or distortions in model comparison, leading to more robust and interpretable conclusions. In other words, when multiple models are considered, maintaining prior compatibility helps preserve coherence in the inference process, ensuring that differences in posterior distributions reflect genuine model distinctions rather than inconsistencies in prior specification.

4.2 Bayesian Analysis of Regression Graphs

The exploration of Bayesian analysis within regression graph models has remained relatively limited, with a few notable contributions, such as those by Ntzoufras and Tarantola (2013) and Ntzoufras et al. (2019). These papers specifically address graphical models of marginal independence using the marginal log-linear parameterization illustrated in Chap. 2. The current literature primarily focuses on Bayesian analysis of discrete regression graph models that are Markov equivalent to bi-directed graphs and DAGs, drawing upon established studies. In situations where this equivalence does not exist, we conjecture that the available methodology could be also employed with suitable generalizations. Further insights and potential developments in this area will be discussed in Sect. 4.6.

Despite focusing on a subset of regression graphs, significant limitations remain. A critical aspect is that, with few exceptions, the probability parameters of a regression graph model cannot be analytically derived in terms of the log-linear parameters introduced in Chap. 2. While this is not problematic within a frequentist approach, it poses significant challenges from a Bayesian perspective. Specifically, conjugate analysis is feasible only for certain graphical configurations, otherwise computational methods, for instance based on MCMC techniques, are required to perform posterior inference. Another complication is that variation independence for the parameter space is not always assured, which can lead to potential inconsistencies and significant challenges in the implementation of MCMC.

In the following, the Bayesian analysis will be pursued using two distinct methods: one revolves around cell probabilities, while the other centers on the use of marginal log-linear parameters. The first method offers the advantage that, under certain models of conditional independence, the likelihood can be directly formulated in terms of probability parameters. In this way, both conjugate and conditional conjugate Bayesian analyses can be carried out effectively.

In contrast, the second approach uses marginal log-linear parameters, which act as clear and interpretable measures of associations among observed variables. Employing a prior distribution on these marginal log-linear interactions, instead of on the probability parameters, proves particularly useful when there is prior knowledge about the odds of specific marginal associations. This approach facilitates the incorporation of symmetry constraints, the elimination of high-order associations, or the inclusion of additional prior information regarding joint and marginal distributions. This is achieved by setting linear constraints on marginal log-linear terms rather than imposing non-linear, multiplicative constraints in the probability space. However, model selection remains a significant challenge in this context and further research is needed.

4.3 Markov Equivalence

There are situations in which two regression chain graphs have exactly the same conditional independencies, a property called *Markov equivalence*. As bi-directed graphs and DAGs are both special cases of regression chain graphs it can happen that a regression chain graph has the same independences of a bi-directed graph or of a DAG.

This property can be verified using a result due to Wermuth and Sadeghi (2012) according to which *two regression graphs are Markov equivalent if and only if they have the same skeleton and the same sets of collision Vs, irrespective of the type of edge.* By the *skeleton* of a regression graph we mean an undirected graph obtained

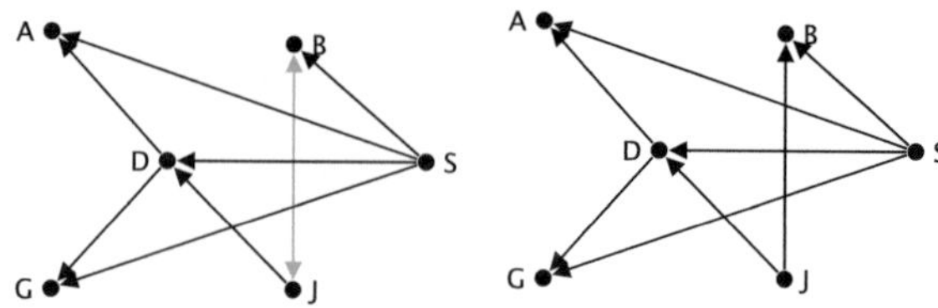

Fig. 4.1 A regression graph and a Markov equivalent DAG

by replacing each edge by a full line. Moreover the *collision* Vs for a regression graph are subgraphs with three nodes and two edges of the following types:

that generalize the definition used for the DAGs. The collision Vs can be compactly denoted by $\bullet \diamond \longrightarrow \bullet \longleftarrow \diamond \bullet$, where the symbol $\diamond$ can be replaced by an arrow tip or by nothing. The inner node is sometimes named an *unshielded collider*. To give an illustration of Markov equivalence between regression graphs and DAGs, let us discuss some regression graphs for the analysis of GSS data; some of them have been used in Sect. 3.4.

Example 4.2 Consider the regression graph for GSS data in Fig. 4.1; it is shown to be Markov equivalent to a DAG that can be obtained by replacing the arc $J \leftrightarrow B$ with a directed edge $J \to B$ since $J \to B \leftarrow S$ is still an unshielded collider and does not modifies the remaining collision Vs.

A second regression graph is considered in Fig. 4.2 (left) obtained by the inclusion of an additional bi-directed edge $D \leftrightarrow G$ in the first block. This graph is not Markov equivalent to a DAG because the only collision V that can be changed by substituting an arc with an arrow is $J \leftrightarrow B \leftarrow S$ but is not enough to obtain a DAG but only a Markov equivalent regression graph shown in Fig. 4.2 (right). $\square$

As proved in Wermuth and Sadeghi (2012), a regression graph can be oriented to achieve Markov equivalence with a DAG possessing the same skeleton if, and only if, it lacks any chordless collision path involving four vertices. This condition is not commonly met. Apart from the example in Fig. 4.1, it is not feasible to orient the regression graphs discussed in Chaps. 2 and 3 used in the analysis of GSS data to be Markov equivalent to a DAG. This limitation primarily arises because regression graph models address certain limitations of DAGs, such as their inability to model two or more response variables jointly.

We now shift our focus to discuss the Markov equivalence between regression graphs and bi-directed graphs.

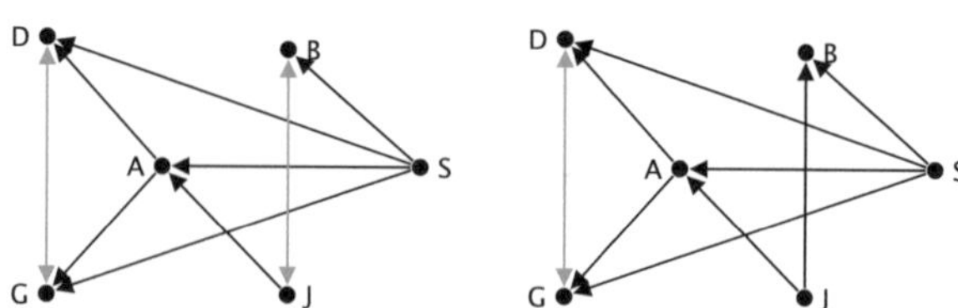

Fig. 4.2 Two Markov equivalent regression graphs without any Markov equivalent DAG

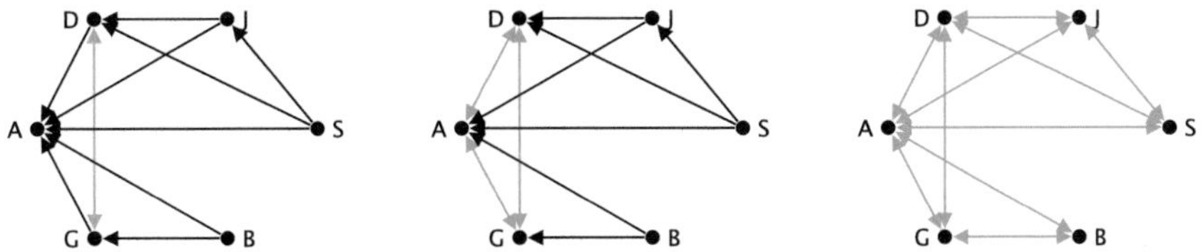

Fig. 4.3 Two Markov equivalent regression graphs and a Markov equivalent bi-directed graph

Example 4.3 Consider the two regression graphs in Fig. 4.3; they define the same independence model for GSS data, even if with a different block configuration. The graph on the left panel considers opinion A as a final response and opinions D, G as a joint intermediate response variable while the graph in the middle models opinions A, D, G as a joint trivariate response. Both regression graphs are Markov equivalent to the bi-directed graph with the same skeleton on the right panel where all arrows are replaced by bi-directed arcs. The bi-directed graph considers all variables on an equal stage and any ordering is compatible with the graph. On the other hand, in the regression graph representations, variables are preliminary grouped in blocks and then ordered according to subject-matter considerations. Even if the interpretation of the statistical models represented by the three graphs is quite different, they indeed define the same probability model as they encode the same set of independence constraints on the joint distribution. $\square$

More generally, a chordless collision path of bi-directed edges is always Markov equivalent to a regression graph where the extreme bi-directed edges are replaced by two arrows preserving the collision set. For instance, a chordless collision path of 4 vertices, must have three edges

Thus the middle edge is always a bi-directed arc and the other two edges $\diamond\!\longrightarrow$ can be an arrow $\longrightarrow$ or an arc $\longleftrightarrow$. Figure 4.4 is illustrative of this property. The Markov equivalence between bi-directed and regression graphs, when feasible, will be employed for Bayesian inference purposes.

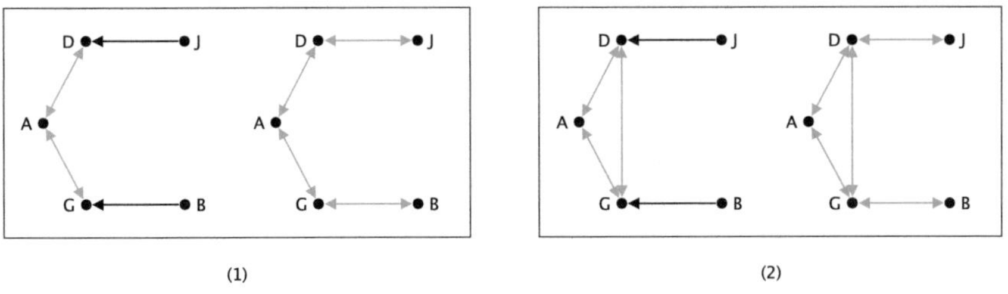

Fig. 4.4 Two examples of Markov equivalence between regression and bi-directed graphs

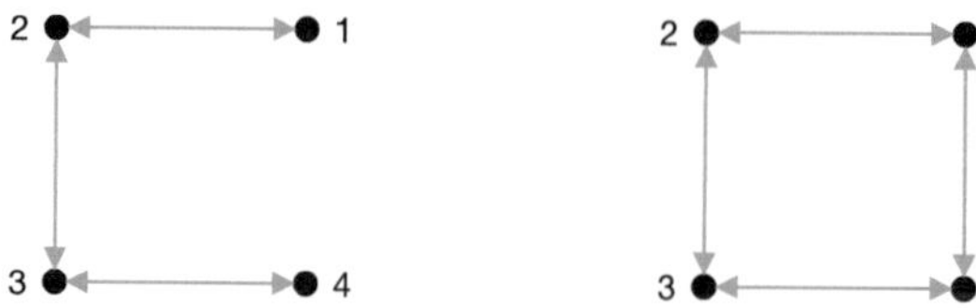

Fig. 4.5 Bi-directed 4-chain (left) and bi-directed chordless 4-cycle (right)

4.4 Cell Probability Parameterization

Under this approach both the prior distributions and the proposals are expressed in terms of the probability parameters. Once the joint posterior distribution of the probability parameters is obtained, the corresponding posterior distribution of the marginal log-linear parameters can be obtained by using the transformation equation (2.21) and then applying Monte Carlo simulations.

We focus on regression graphs that are Markov equivalent to a bi-directed graph. We distinguish between homogeneous and non-homogeneous bi-directed graphs. A bi-directed graph is considered homogeneous if it lacks a bi-directed 4-chain or a chordless 4-cycle within its subgraphs, see Fig. 4.5.

Example 4.4 The bi-directed graphs of Fig. 4.4 are both non-homogeneous. The one in panel (1) is a 5-chain, and the one in panel (2) includes the subgraph with vertex set $\{J, D, G, B\}$ which is a 4-chain. Also, the bi-directed graph in the right-hand side of Fig. 4.3 is non-homogeneo us since it contains the 4-chain $\{B, G, D, S\}$. $\square$

Both homogeneous and non-homogeneous models are shown to be compatible, in terms of independencies, with a certain DAG representation. While homogeneous models can be depicted using a DAG with the same set of vertices, representing non-homogeneous models via a DAG requires incorporating additional latent variables; in the following, for simplicity, we will use the expression *augmented* DAG to denote both types of DAGs.

To construct an augmented DAG, we follow the methodology outlined by Pearl and Wermuth (1994) and previously discussed in Sect. 1.7. Starting with the skeleton

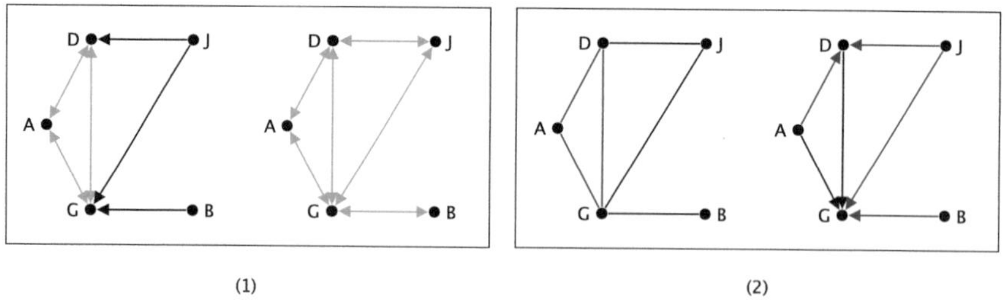

(1)   (2)

Fig. 4.6 **1** Regression graph and Markov equivalent bi-directed graph. **2** Skeleton and Augmented DAG representation

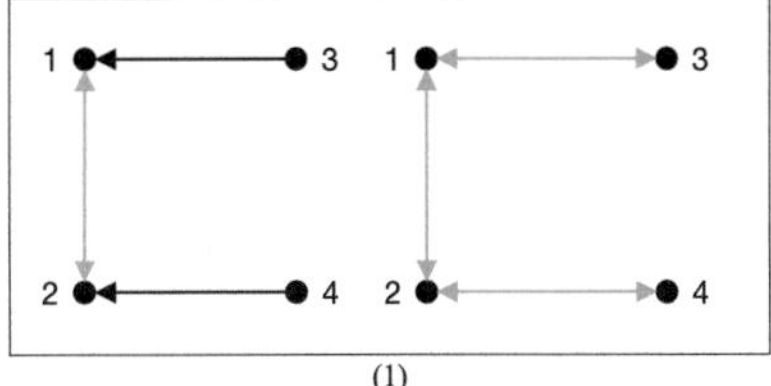
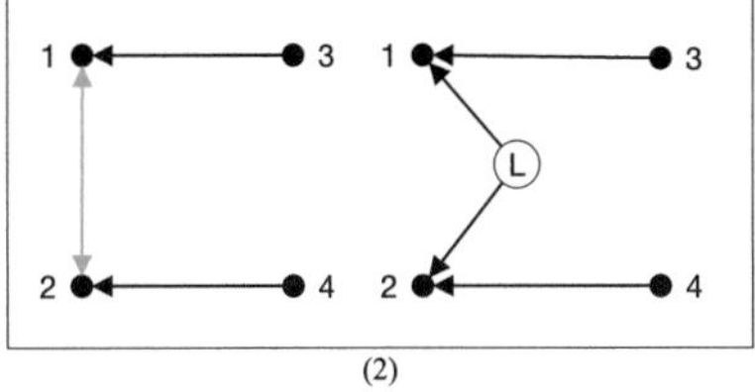

Fig. 4.7 **1** Regression graph and Markov equivalent bi-directed graph. **2** Sink orientation and Augmented DAG representation

$\overline{\mathcal{G}}$ of the target graph, we establish a *sink* orientation by assigning directional arrows $v_i \longrightarrow v_j \longleftarrow v_k$ to every chordless 3-chain in $\overline{\mathcal{G}}$, such as $v_i - v_j - v_k$. Then, if the resulting sink orientation shows some bi-directed edges, each of them is replaced by a directed structure $v_i \longleftarrow L \longrightarrow v_j$, where L denotes a latent variable, leading to a graph with additional latent nodes. Finally, a DAG is formed by assigning acyclic orientations to undirected edges in this sink orientation. The resulting DAG $\mathcal{D}_G$, referred to as the augmented DAG of $\mathcal{G}$, incorporates latent variables in the case of non-homogeneous models to account for each bi-directed edge identified in the sink orientation, while homogeneous models' augmented DAGs do not require latent variables. The augmented vertex set is denoted with $\mathcal{A} = \mathcal{V} \cup \mathcal{L}$, where $\mathcal{L}$ is the set of additional latent variables; if $\mathcal{G}$ is homogeneous then $\mathcal{L} = \emptyset$ and $\mathcal{A} = \mathcal{V}$.

Example 4.5 The Wermuth and Pearl procedure is illustrated in Fig. 4.6. The left panel shows a regression graph alongside its corresponding Markov-equivalent bi-directed graph, which is homogeneous. On the right-hand panel we represent its skeleton with undirected edges and the corresponding sink DAG representation. $\square$

Example 4.6 Figure 4.7 considers instead the situation of a non-homogeneous bi-directed graph. The left panel represents the regression graph and the Markov equivalent bi-directed graph. Since the sink orientation contains a bi-directed edge a latent variable is added in the right panel to create the corresponding augmented DAG. $\square$

In the sequel, models associated with bi-directed graphs are denoted using the conventional notation for undirected graphs. These models are uniquely characterized by their maximal cliques, that is the largest subsets of nodes that induce complete subgraphs. For instance, given the bi-directed graph in Fig. 4.6 (1) with maximal cliques $\{\{A, D, G, \}, \{D, G, J\}, B\}$, the independence model is defined as $\sim A * D * G + D * G * J + B$, as well as the bi-directed graph model in Fig. 4.7(1) is defined as $\sim 1 * 3 + 1 * 2 + 2 * 4$.

4.4.1 Prior Distribution

By using the augmented DAG representation, we can parameterize the model via a minimal set of marginal and conditional probability parameter vector $\boldsymbol{\Pi}$ sufficient to obtain the joint distribution of interest. In the sequel, the indices i of the contingency table cell will be used extensively. To keep the notation simpler, boldface will no longer be employed. The set Π is defined as

$$\Pi = \mathrm{vec}\left(\pi_{v|\mathrm{pa}(v)}(i_v^*|i_{\mathrm{pa}(v)}^*);\ i_v^* \in \mathcal{I}_v \setminus \left\{|\mathcal{I}_v|\right\},\ i_{\mathrm{pa}(v)}^* \in \mathcal{I}_{\mathrm{pa}(v)},\ v \in \mathcal{A}\right).$$

The vector of joint probabilities $\pi_{\mathcal{A}}$ corresponding to the augmented set of variables $X_{\mathcal{A}}$ can be expressed trough the conventional DAG factorization method as follows

$$\pi_{\mathcal{A}}(i^*) = \prod_{v \in \mathcal{A}} \pi_{v|\mathrm{pa}(v)}\left(i_v^*|i_{\mathrm{pa}(v)}^*\right),\ \text{for } i \in \mathcal{I}_{\mathcal{V}}, \tag{4.7}$$

where $\mathrm{pa}(v)$ stands for the parents set of vertex v in graph $\mathcal{D}_{\mathcal{G}}$ and $\pi_{v|U}(i_v|i_U)$ is the parameter for the conditional probability $P(X_v = i_v|X_U = i_U)$.

The corresponding joint probabilities associated to the observable variables $X_{\mathcal{V}}$ are obtained from $\pi_{\mathcal{A}}$ summing over the levels of the latent variables, that is,

$$\pi(i) = \sum_{i_{\mathcal{L}} \in \mathcal{I}_{\mathcal{L}}} \pi_{\mathcal{A}}(i, i_{\mathcal{L}})\ \text{for } i \in \mathcal{I}_{\mathcal{V}}.$$

The augmented likelihood for a specific $\mathcal{D}_{\mathcal{G}}$ is given by

$$f(n^{\mathcal{A}}|\Pi) = \frac{\Gamma(N+1)}{\displaystyle\prod_{i^* \in \mathcal{I}_{\mathcal{A}}} \Gamma(n^{\mathcal{A}}(i_{\mathcal{A}}) + 1)} \prod_{v \in \mathcal{A}} \left\{\prod_{i_{\mathrm{cl}(v)}^* \in \mathcal{I}_{\mathrm{cl}(v)}} \pi_{v|\mathrm{pa}(v)}\left(i_v|i_{\mathrm{pa}(v)}\right)^{n^{\mathcal{A}}(i_{\mathrm{cl}(v)})}\right\},$$

$$\tag{4.8}$$

where $\mathrm{cl}(v) = \{v\} \cup \mathrm{pa}(v)$ is the closure of the vertex v and $n^{\mathcal{A}} = \left(n^{\mathcal{A}}(i_{\mathcal{A}}), i_{\mathcal{A}} \in \mathcal{I}_{\mathcal{A}}\right)$ is the vector of the cell frequencies of the augmented contingency table for the variables $\mathcal{A}$. In the scenario where the bi-directed graph is homogeneous, the DAG representation of $\mathcal{G}$ excludes any latent variables, thereby rendering the model likelihood directly as given by (4.8).

This setup enables the use of a product of Dirichlet distributions exploiting the procedure developed for DAG models. In the case of homogeneous models, these prior distributions are conjugate to the model likelihood, thereby facilitating the inference process. Conversely, for non-homogeneous models, inference is performed via a conditional conjugate approach using the augmented DAG; the posterior distribution of the joint probabilities is derived by summing over all levels of the latent variables.

The prior distribution on the probability parameter vector can be factorized according to the structure of the corresponding Markov equivalent DAG as follows

$$f(\Pi) = \prod_{v \in \mathcal{A}} \prod_{i_{\mathrm{pa}(v)}^* \in \mathcal{I}_{\mathrm{pa}(v)}} f_{\mathcal{D}i}\left(\pi_{v|i_{\mathrm{pa}(v)}};\ \alpha_{v|i_{\mathrm{pa}(v)}}\right), \tag{4.9}$$

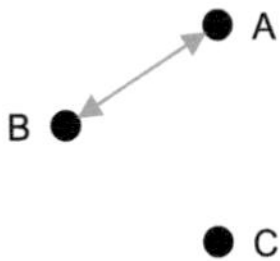

Fig. 4.8 A two component bi-directed graph

where $\alpha_{v|i_{\mathrm{pa}(v)}} = \left(\alpha_{\mathrm{cl}(v)} \left(i_{\mathrm{cl}(v)}\right)\right); i_v \in \mathcal{I}_v$ and $f_{\mathcal{D}i}(\pi; \alpha)$ is the Dirichlet density function with parameters α evaluated at π.

Example 4.7 For the graph in Fig. 4.8 the prior will be a product of a Dirichlet distributions for π_{AB} and π_C with parameters α_{AB} and α_C respectively.

$$f(\Pi) = f_{\mathcal{D}i}\left(\pi_{A|i_B}; \alpha_{A|i_B}\right) \times f_{\mathcal{D}i}\left(\pi_B; \alpha_B\right) \times f_{\mathcal{D}i}\left(\pi_C; \alpha_C\right)$$

$\square$

As discussed in Sect. 4.1, compatibility on prior specification is an important aspect to consider. In order to have compatibility across models, we assign a Dirichlet distribution on the vector of joint probabilities for the saturated model of the observed table. We proceed in a similar way for non-homogeneous models working in terms of the saturated augmented DAG, and we ensure that such prior on π is the same as the one considered initially. We obtain compatibility setting $\alpha(i) = \sum_{i_\ell \in I_L} \alpha_A(i, i_\ell)$ and we derive the hyperparameters of each element in (4.9) as $\alpha_{\mathrm{cl}(v)}\left(i_{\mathrm{cl}(v)}\right) = \sum_{i_{A \setminus \mathrm{cl}(v)} \in I_{A \setminus \mathrm{cl}(v)}} \alpha_A\left(i_{\mathrm{cl}(v)}, i_{A \setminus \mathrm{cl}(v)}\right)$. The choice of this prior parameter value is of prominent importance for the model comparison due to the well-known sensitivity of the posterior model odds and the Bartlett-Lindley paradox. Briefly, this paradox occurs when the Bayesian evidence in favor of a model becomes overwhelming with increasingly vague priors, regardless of the actual data observed. In terms of model comparisons this leads to the fact that more parsimonious models are supported irrespective of which data we observe. When no prior information is available, a usual choice is to consider equal $\alpha(i)$ for all cells $i \in \mathcal{I}$. Common choices for $\alpha(i)$ are $1/2$ (Jeffreys prior), 1 (Unit Expected Cell prior, UEC) and $1/|\mathcal{I}|$ (Perks prior). Perks prior has a unit information interpretation and it can be used as a yardstick in order to identify and interpret the effect of any other prior distribution used. Prior distributions with $\alpha(i) < 1/|I|$ result in larger variance than the one imposed by our proposed unit information prior and hence they a posteriori support more parsimonious models. On the contrary, prior distributions with $\alpha(i) > 1/|I|$ result in lower prior variance and hence they a posteriori support models with more complicated graph structure. A comparison between the previously examined prior distributions is provided in Table 4.1.

Table 4.1 Table of prior variance in comparison to Perk's prior (last row of the table)

	Parameter	$V\big[\pi(i)\big]$	Variance ratio
Jeffrey's	$\alpha(i) = 1/2$	$2\,\dfrac{|\mathcal{I}|-1}{|\mathcal{I}|^2\{|\mathcal{I}|+2\}}$	$4/\big(|\mathcal{I}|+2\big)$
Unit Expected Cell (UEC)	$\alpha(i) = 1$	$\dfrac{|\mathcal{I}|-1}{|\mathcal{I}|^2\{|\mathcal{I}|+1\}}$	$2/\big(|\mathcal{I}|+1\big)$
Perks' Prior (UIP-Perks')	$\alpha(i) = 1/|\mathcal{I}|$	$\dfrac{|\mathcal{I}|-1}{2|\mathcal{I}|^2}$	1

From (4.8) and (4.9), the posterior distribution of the parameters $\boldsymbol{\Pi}$ given a set of augmented data $\boldsymbol{n}^{\mathcal{A}}$ is given by

$$f\left(\boldsymbol{\Pi}|\boldsymbol{n}^{\mathcal{A}}\right) = \prod_{v\in\mathcal{A}}\ \prod_{i_{\mathrm{pa}(v)}\in\mathcal{I}_{\mathrm{pa}(v)}} f_{\mathcal{D}i}\left(\boldsymbol{\pi}_{v|i_{\mathrm{pa}(v)}};\ \widetilde{\boldsymbol{\alpha}}_{v|i_{\mathrm{pa}(v)}}\right), \qquad (4.10)$$

where $\widetilde{\boldsymbol{\alpha}}_{v|i_{\mathrm{pa}(v)}} = \left(\widetilde{\alpha}_{\mathrm{cl}(v)}(i_{\mathrm{cl}(v)}) = n^{\mathcal{A}}_{\mathrm{cl}(v)}(i_{\mathrm{cl}(v)}) + \alpha_{\mathrm{cl}(v)}(i_{\mathrm{cl}(v)}),\ i_v \in \mathcal{I}_v\right)$, for any given configuration $i_{\mathrm{pa}(v)} \in \mathcal{I}_{\mathrm{pa}(v)}$.

In order to obtain a sample from the posterior distribution of $\boldsymbol{\Pi}$ we consider a Gibbs algorithm consisting of the following two steps

Step 1:   Generate frequencies of the augmented table: use a Multinomial distribution to generate frequencies for an augmented table.

Step 2:   Sample from conditional distributions: Generating random samples from the conditional Dirichlet posterior distributions for each set of probability parameters of the augmented DAG, given the augmented table from Step 1.

Additional constraints are imposed on model parameters to remain in the class of identifiable models; see Ntzoufras and Tarantola (2013) for the details. The joint and the marginal probabilities for the observed variables are obtained by implementing the appropriate marginalizing transformations to the probability parameter values of Step 2. Additionally, the posterior distribution of the log-linear interactions can be obtained as a by-product of this approach This can be achieved by calculating appropriate contrasts of the logarithms of the joint probability parameters obtained at each single iteration. Zero constraints on the λ parameters are automatically imposed by construction.

4.4.2   *Model Selection*

As discussed in Sect. 4.1, for model choice we need to estimate the posterior model probabilities $f(\mathcal{G}|n) \propto f(\boldsymbol{n}|\mathcal{G})f(\mathcal{G})$, with $f(\boldsymbol{n}|\mathcal{G})$ marginal likelihood of the model and $f(\mathcal{G})$ prior distribution on $\mathcal{G}$. Among the alternative models we will select the one with the highest posterior probability. In the absence of prior information, a uniform prior on G will be assumed and the posterior will depend only on the marginal likelihood $f(\boldsymbol{n}|\mathcal{G})$ of the model under consideration.

For homogeneous models, the marginal likelihood can be analytically calculated as follows

$$f(\boldsymbol{n}|\mathcal{G}) = \mathrm{Co}(\boldsymbol{n}) \times \prod_{v \in \mathcal{V}} \prod_{i_{\mathrm{pa}(v)} \in \mathcal{I}_{\mathrm{pa}(v)}} \frac{\mathrm{MB}\big(\widetilde{\boldsymbol{\alpha}}_{v|i_{\mathrm{pa}(v)}}\big)}{\mathrm{MB}\big(\boldsymbol{\alpha}_{v|i_{\mathrm{pa}(v)}}\big)}$$

where

$$\mathrm{Co}(\boldsymbol{n}) = \frac{\Gamma(N+1)}{\prod_{i \in \mathcal{I}} \Gamma\big(n(i_{\mathcal{V}}) + 1\big)}$$

is the multinomial constant and

$$\mathrm{MB}(\boldsymbol{\alpha}) = \frac{\prod_{i \in \mathcal{I}} \Gamma\big(\alpha(i)\big)}{\Gamma\left(\sum_{i \in \mathcal{I}} \alpha(i)\right)}$$

is the normalizing constant of the multinomial beta function, that is the marginal likelihood of the multinomial distribution when using conjugate Dirichlet prior.

On the other hand for non-homogeneous models the marginal likelihood is obtained via Chib's estimator, Chib (1995). An estimate of the marginal likelihood is given by

$$\widehat{f}(\boldsymbol{n}|\mathcal{D}) = \frac{f(\boldsymbol{n}|\boldsymbol{\pi}^{*\mathcal{D}})f(\boldsymbol{\pi}^{*\mathcal{D}})}{f(\boldsymbol{\pi}^{*\mathcal{D}}|\boldsymbol{n})}, \tag{4.11}$$

where $\boldsymbol{\pi}^{*\mathcal{D}}$ corresponds to the mode of the distribution. The use of an high posterior density point ensures to get reliable estimates. A correction for label switching is implemented in such a way as to obtain a posterior distribution that is invariant to permutations in the labelling of the parameters The Bayes Factor (BF) for comparing models $\mathcal{G}_1$ and $\mathcal{G}_2$ is obtained in closed form and given by

$$\mathrm{BF} = \frac{f(\boldsymbol{n}|\mathcal{G}_1)}{f(\boldsymbol{n}|\mathcal{G}_2)}.$$

4.4.3 Bayesian Inference for Coppen's Data

The data discussed in Example 2.6 concern 4 variables, leading to 64 potential bi-directed graphs. It can be shown that 13 of them exhibit non-homogeneity comprising 12 chains and only one single cycle.

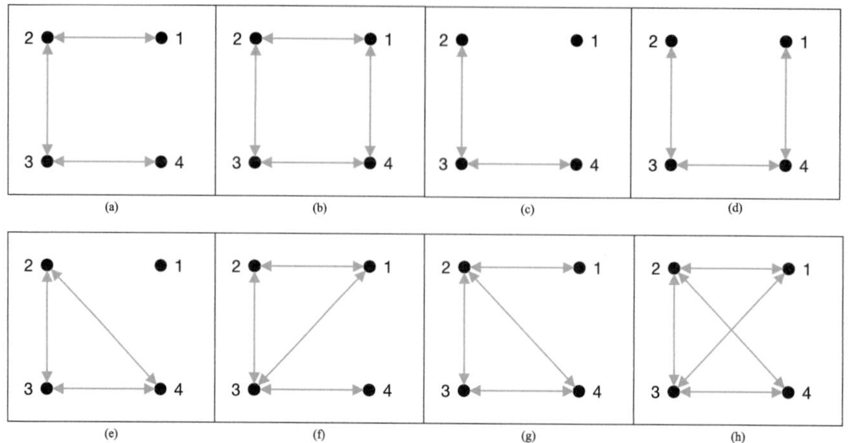

Fig. 4.9 Bi-directed graph models for the analysis of Coppen's data in Table 4.2

In Table 4.2 we present all models with estimated posterior probability higher than 0.01 under the three examined prior set-ups (Perks prior, Jeffreys prior, and unit expected cells prior). For all models we report the batch mean estimate, its standard error and the standard deviation of the marginal log-likelihood and the corresponding posterior model probabilities over 30 MCMC samples. In all simulations we used 3000 iterations for the bi-directed four chains and 10,000 iterations for the chordless four cycle. The batch mean estimate is a technique used to assess the accuracy of parameter estimates derived from MCMC simulations. The entire chain of MCMC samples is divided into several equal-sized segments or "batches" (in this case, 30). For each batch, the mean is calculated, providing a representative value for that segment. Averaging these means across all batches yields an overall estimate of the parameter. This method minimizes the impact of sample autocorrelation, thus improving the estimate reliability. The standard error of the batch mean is calculated by measuring the variability across batches, helping to gauge the precision of the final parameter estimate.

Figure 4.9 offers a visual representation of the bi-directed graphs examined in Table 4.2. Except for graph (b), all the bi-directed graphs are shown to be Markov-equivalent to a regression graph, as detailed in Fig. 4.10. These Markov equivalent graphs are derived following the rule outlined in Sect. 4.3. The same rule shows that the independence model encoded by the bi-directed graph (b), i.e., $X_2 \perp\!\!\!\perp X_4$, $X_1 \perp\!\!\!\perp X_3$, cannot be represented by any regression graph. It is worth noticing that as far as the bi-directed graph (d) is concerned, the Markov equivalent regression graph assumes a different grouping of variables in chain components: unlike other regression graphs, the response component includes nodes $\{3, 4\}$ and nodes $\{1, 2\}$ are collected in the background component.

The maximum a posteriori (MAP) model , under all prior set-ups, is the graph (a) with probabilities 0.67, 0.91 and 0.42 for Perks, Jeffreys and UEC prior, respectively.

Table 4.2 Marginal log-likelihood and posterior probabilities (% values) for models with estimated posterior probability >0.01 for Coppen's data (the batch mean estimates, standard errors (se), and standard deviations (std) over 30 samples are reported); 3,000 and 10,000 iterations were used for the 4-chain and the chordless 4-cycle bi-directed graphs respectively

Perks Prior $\alpha(i) = 1/2^4$

		Marginal log-likelihood		Posterior probability (%)	
Rank	Model	Mean (se)	S.D.	Mean (se)	Std.
1	$\sim 1*2 + 2*3 + 3*4$ (a)	-64.75 (0.095)	0.523	67.31 (4.791)	26.24
2	$\sim 1*2 + 2*3 + 3*4 + 1*4$ (b)	-66.21 (0.370)	2.029	30.85 (4.893)	26.80
3	$\sim 1 + 2*3 + 3*4$ (c)	-68.97 (0.000)	0.000	1.04 (0.113)	0.62
4	$\sim 1*4 + 2*3 + 3*4$ (d)	-69.89 (0.083)	0.457	0.49 (0.095)	0.52
5	$\sim 1 + 2*3*4$ (e)	-71.10 (0.000)	0.000	0.12 (0.013)	0.07

Jeffreys Prior $\alpha(i) = 1/2$

		Marginal log-likelihood		Posterior probability (%)	
Rank	Model	Mean (se)	Std.	Mean (se)	Std.
1	$\sim 1*2 + 2*3 + 3*4$ (a)	-56.74 (0.030)	0.165	91.06 (0.269)	1.47
2	$\sim 1 + 2*3 + 3*4$ (c)	-59.79 (0.000)	0.000	4.37 (0.126)	0.69
3	$\sim 1 + 2*3*4$ (e)	-60.44 (0.000)	0.000	2.27 (0.065)	0.36
4	$\sim 1*4 + 2*3 + 3*4$ (d)	-61.55 (0.040)	0.219	0.77 (0.044)	0.24
5	$\sim 1*2*3 + 3*4$ (f)	-61.61 (0.000)	0.000	0.70 (0.020)	0.11

Unit Expected Cell Prior $\alpha(i) = 1$

		Marginal log-likelihood		Posterior probability (%)	
Rank	Model	Mean (se)	Std.	Mean (se)	Std.
1	$\sim 1*2 + 2*3 + 3*4$ (a)	-56.68 (0.012)	0.068	42.57 (0.303)	1.66
2	$\sim 1*2*3 + 3*4$ (f)	-57.59 (0.000)	0.000	17.14 (0.089)	0.48
3	$\sim 1 + 2*3 + 3*4$ (c)	-57.81 (0.000)	0.000	13.76 (0.071)	0.39
4	$\sim 1 + 2*3*4$ (e)	-58.11 (0.000)	0.000	10.20 (0.053)	0.29
5	$\sim 1*2 + 2*3*4$ (g)	-58.56 (0.000)	0.000	6.55 (0.034)	0.19
6	$\sim 1*2*3 + 2*3*4$ (h)	-59.24 (0.000)	0.000	3.31 (0.017)	0.09

For the Perks prior, the relative difference between the MAP model and the second best graph (b) is smaller. Even if we increase 100 000 the number of iterations for graph (b) the Monte Carlo variability remains quite high (marginal log-likelihood Monte Carlo error ≈ 1.24).

In order to summarize the previous results in a single representative graph we can construct the Median Probability Model, that is the graph obtained by including edges with posterior inclusion probabilities equal or >0.5, The situation is summarized in Table 4.3 According to this table, the edges $1 \longleftrightarrow 2$, $2 \longleftrightarrow 3$ and $3 \longleftrightarrow 4$ should be included in the final selected graph with posterior inclusion probabilities, under the considered prior set-ups, at least 0.73, 1.00 and 0.99, respectively. The bi-directed 4-chain graph (a) is the median probability model under all prior set-ups.

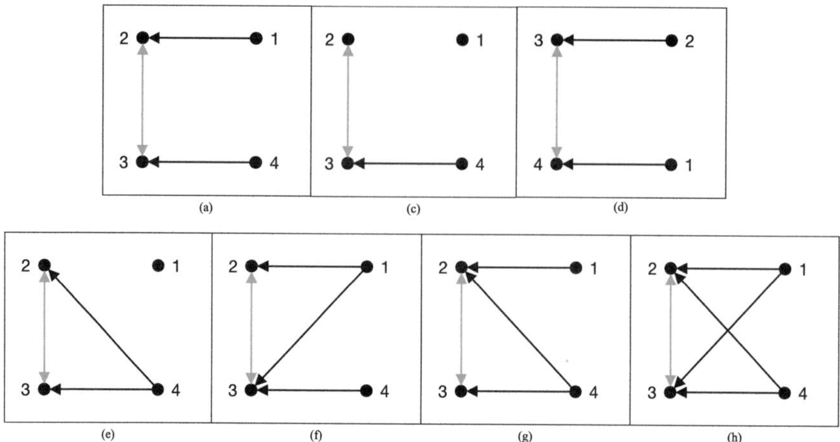

Fig. 4.10 Regression graphs that are Markov equivalent to bi-directed graphs in Fig. 4.9

As a by-product we can estimate the posterior distribution of λ using Monte Carlo samples from the posterior distribution of π^A. Specifically, a sample from the posterior distribution of λ can be generated by the following iterative procedure. At each iteration t (for $t = 1, \ldots, T$):

(i)   Generate a random value $\pi^{A(t)}$ from the posterior distribution of π^D.
(ii) Calculate the full table of probabilities $\pi^{(t)}$ from $\pi^{A(t)}$.
(iii) Obtain marginal log-linear parameter vector $\lambda^{(t)}$ from $\pi^{(t)}$ via Eq. (2.21).

Figures 4.11 and 4.12 represent posterior distributions for the joint probabilities π and marginal log-linear parameters λ for the MAP bi-directed graph model in Fig. 4.9a equivalent to the regression graph in Fig. 4.10a. Posterior values are obtained

Table 4.3 Posterior inclusion probabilities (% values) for each edge of the bi-directed 4-way graph for Coppen's data (the batch mean estimates (standard errors) over 30 samples are reported; 3,000 and 10,000 iterations were used for the 4-chain and the chordless 4-cycle bi-directed graphs respectively)

| | Prior set-up | | |
| | Perks | Jeffreys | UEC |
Edge	$\alpha(i) = 1/16$	$\alpha(i) = 1/2$	$\alpha(i) = 1$
12	98.2 (0.22)	92.5 (0.22)	73.8 (0.14)
13	0.0 (0.00)	0.9 (0.03)	23.8 (0.12)
14	31.4 (4.85)	0.9 (0.06)	4.3 (0.03)
23	100.0 (0.00)	100.0 (0.00)	100.0 (0.00)
24	0.1 (0.01)	2.7 (0.08)	23.1 (0.12)
34	99.8 (0.02)	99.7 (0.01)	99.0 (0.01)

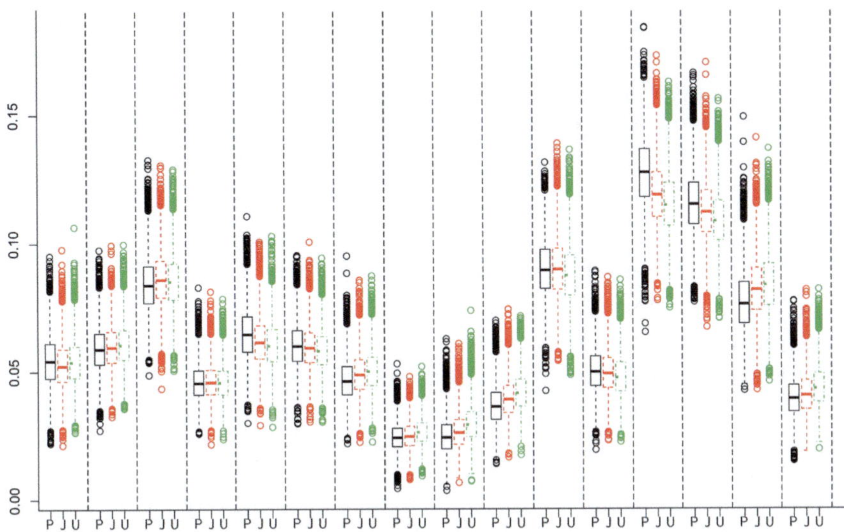

Fig. 4.11 Boxplots of the posterior distribution of the joint probability parameter π for the graphical model in Fig. 4.9a; each block of box-plot refers to elements of the probability vector in reverse lexicographic order

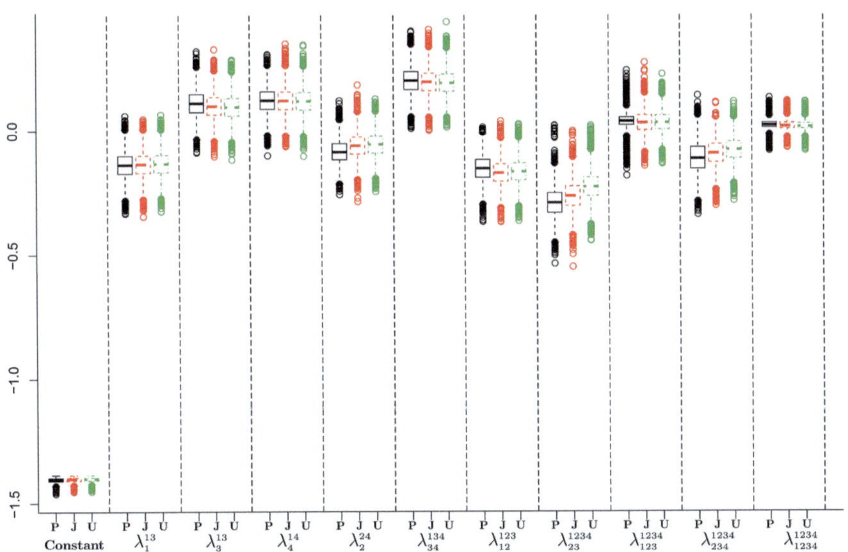

Fig. 4.12 Boxplots of the posterior distributions of the marginal log-linear parameter λ for the graphical model Fig. 4.9a

using 3,000 iterations. We note only minor differences under the examined prior set-ups.

If one aims to conduct inference on the marginal log-linear parameters by directly assigning a prior distribution, the procedure outlined in this section is however not applicable. A potential strategy is presented in the following section.

4.5 Marginal Log-Linear Parameterization

The probability-based method exhibits significant limitations in incorporating a priori knowledge about marginal associations among observed variables. This method inherently imposes non-linear, multiplicative constraints on the probability space, complicating the modeling process and limiting flexibility in the specification of prior knowledge.

In contrast, marginal log-linear parameterization offers a more straightforward and effective approach to specify sub-models based on a priori knowledge. This approach accommodates linear constraints on the marginal log-linear parameter space, providing a more intuitive and mathematically tractable framework for model specification and the integration of prior information. For instance, symmetry constraints, context-specific independence assumptions, or additional prior information about the joint and marginal distributions can be integrated by setting zero or linear constraints on the marginal log-linear parameter space. Consequently, specifying sub-models based on a priori knowledge is markedly simpler with the marginal log-linear parameterization compared to the probability-based approach.

Prior distributions for the marginal log-linear parameters can be elicited following Knuiman and Speed (1988) and Dellaportas and Forster (1999). Given that a conjugate analysis is not feasible, reliance on MCMC methods becomes necessary. Two possible approaches are considered:

- Express both prior and proposal in terms of the marginal log-linear parameters. While this approach is intuitively appealing, it requires iterative methods to calculate the cell probabilities and, consequently, the model likelihood in each iteration of the MCMC sampler, potentially slowing down the algorithm's speed.
- Express the prior in terms of the marginal log-linear parameters, but use proposals on the probability parameters. This strategy involves a parameter transformation to obtain the corresponding proposal on the marginal log-linear interactions, facilitating movement within the desired target space. By exploiting the conditional conjugate setup presented in the previous section, we can construct an efficient proposal mechanism for probability parameters, enhancing the overall efficiency of the MCMC algorithm. Furthermore, in this way we ensure to move within the space of variation independent models.

4.5.1 Prior Distribution

In this section, we delve into the process of prior specification for marginal log-linear models, discussing alternative strategies. We will restrict our attention to the case of variation independent models.

As previously detailed in Eq. (2.21), a bi-directed graph can be modeled by a marginal log-linear parameterization

$$\lambda = C_{\mathcal{M}} \log(L_{\mathcal{M}}\pi)$$

induced by the sequence $\mathcal{M}$ disconnected sets with zero constraints

$$K_{\mathcal{M}} \log\left(L_{\mathcal{M}}\pi\right) = 0 \tag{4.12}$$

on some suitable λ parameters identified by the sub-matrix $K_{\mathcal{M}}$ of $C_{\mathcal{M}}$.

We consider the sum to zero constraints on the marginal log linear that allow an easier prior elicitation. The vector π is obtained by arranging the elements $\pi(i)$ in a reverse lexicographic ordering of the corresponding variable levels, with the level of the first variable changing first.

For subsequent discussions, it will be beneficial to rewrite the model described by Eq. (2.21) in the following extended form

$$\begin{bmatrix} \lambda^{M_1} \\ \vdots \\ \lambda^{M_k} \\ \vdots \\ \lambda^{V} \end{bmatrix} = \mathrm{diag}\begin{bmatrix} C_1, \cdots, C_m, \cdots, C_V \end{bmatrix} \begin{bmatrix} \log \pi^{M_1} \\ \vdots \\ \log \pi_{M_k} \\ \vdots \\ \log \pi_{V} \end{bmatrix}, \tag{4.13}$$

so that the interactions λ^{M_k} for each marginal M_k are derived as

$$\lambda^{M_k} = C_{M_k} \log \pi_{M_k}, \quad \forall M_k \in V.$$

Not all interactions in λ^{M_k} are relevant due to constraints imposed by the graphical structure G and the contrast matrix, leading us to consider only the non-zero elements of λ, denoted $\overrightarrow{\lambda}$. This subset includes

$$\overrightarrow{\lambda} = \bigcup_{M_k \in V} \{\lambda_j^{M_k} : \lambda_j^{M_k} \neq 0, j = 1, \ldots, r_{C_m}\},$$

where r_{C_m} represents the number of rows in C_m.

In the absence of specific information about $\overrightarrow{\lambda}$, we can assign to each element $\overrightarrow{\lambda}_j$ an independent normal prior distribution with a large variance to express ignorance

$$f(\overrightarrow{\lambda}_j) \sim N(0, \sigma_j^2), \quad j = 1, 2, \ldots, d_{\overrightarrow{\lambda}},$$

with $d_{\overrightarrow{\lambda}}$ being the total number of elements in $\overrightarrow{\lambda}$.

A refined approach involves adapting the default prior of Dellaportas and Forster (1999) for standard log-linear of the form $\log \mu = X\theta$. This prior is defined by

$$\theta \sim N\left(\vartheta, 2|\mathcal{I}|(X^\mathsf{T}X)^{-1}\right),$$

where $\vartheta = (\log \bar{n}, 0, \ldots, 0)^\mathsf{T}$ with $\bar{n}$ denoting the average of observed cell counts, and $|\mathcal{I}|$ denotes the total number of cells in the contingency table. This prior facilitates model comparison and fits naturally with log-linear marginal models.

For constructing the prior on $\overrightarrow{\lambda}$, we separately consider each λ^{M_k} from marginal M_k. Let $\lambda_S^{M_k}$ be the parameter vector for the saturated model that can be estimated from marginal M_k. The corresponding Dellaportas and Forster's prior is given by

$$\lambda_S^{M_k} \sim N\left(\vartheta - \log(N)X_{M_k}^{-1}\mathbf{1}, 2|\mathcal{I}_{M_k}|(X_{M_k}^\mathsf{T}X_{M_k})^{-1}\right). \tag{4.14}$$

Using sum-to-zero constraints simplifies the prior to independent normals for each interaction, with $\vartheta_j = 0$ for all but the intercept. The compatible prior for each element $\overrightarrow{\lambda}^{M_k}$ is obtained via marginalization from Eq. (4.14) and the full $\overrightarrow{\lambda}$ prior emerges as the product of these individual priors.

Since we are not in a conjugate setting, the posterior distribution of $\overrightarrow{\lambda}$ cannot be obtained in closed form. Next subsection provides an MCMC based algorithm to performe posterior inference.

4.5.2 MCMC Algorithm

In order to obtain the posterior distribution of $\overrightarrow{\lambda}$, we preliminary need to consider the posterior distribution of the augmented set $\lambda^\mathcal{A}$ that is

$$f(\lambda^\mathcal{A}|n) \propto f\left(n|\wp(\lambda)\right)f(\overrightarrow{\lambda})f(\xi), \tag{4.15}$$

where $\wp(\cdot)$ defines the inverse mapping $\pi \mapsto \lambda$ and $f(\xi)$ is a pseudo prior used for the additional parameters included in $\lambda^\mathcal{A}$; $\lambda^\mathcal{A} = (\overrightarrow{\lambda}, \xi)$.

A Metropolis-Hastings (MH) algorithm is then used to sample from the distribution in Eq. (4.15) so to derive the posterior distribution distribution of $\overrightarrow{\lambda}$ via marginalization. The MH algorithm is based on the following steps.

1. Propose a new vector Π' from $q(\Pi'|\Pi^{(t)})$; where $\Pi^{(t)}$ are the values of Π at t iteration.

2. From $\mathbf{\Pi}'$, calculate the proposed joint probabilities $\boldsymbol{\pi}'$ (for the observed table) via marginalization.
3. From $\boldsymbol{\pi}'$, calculate $\boldsymbol{\lambda}'$ using (2.21) and then obtain the corresponding non-zero elements $\vec{\boldsymbol{\lambda}}'$.
4. Set $\boldsymbol{\xi}' = \mathbf{\Pi}'_{\xi}$; where $\mathbf{\Pi}'_{\xi}$ is a pre-specified subset of $\mathbf{\Pi}'$ of dimension $d_{\xi} = \dim(\mathbf{\Pi}) - \dim(\vec{\boldsymbol{\lambda}})$; in our implementation we have used as $\boldsymbol{\xi}$ the last d_{ξ} terms of $\mathbf{\Pi}_{\xi}$.
5. Accept the proposed move with probability $\alpha = \min(1, A)$ with

$$
A = \frac{f(\boldsymbol{n}|\mathbf{\Pi}')f(\vec{\boldsymbol{\lambda}}')f(\boldsymbol{\xi}')q(\mathbf{\Pi}^{(t)}|\mathbf{\Pi}')}{f(\boldsymbol{n}|\mathbf{\Pi}^{(t)})f(\vec{\boldsymbol{\lambda}}^{(t)})f(\boldsymbol{\xi})q(\mathbf{\Pi}'|\mathbf{\Pi}^{(t)})} \times \mathrm{abs}\left(\frac{\mathcal{J}\left(\mathbf{\Pi}^{(t)}, \vec{\boldsymbol{\lambda}}^{(t)}, \boldsymbol{\xi}^{(t)}\right)}{\mathcal{J}\left(\mathbf{\Pi}', \vec{\boldsymbol{\lambda}}', \boldsymbol{\xi}'\right)}\right)
$$

$$(4.16)$$

where $\mathrm{abs}(\cdot)$ stands for the absolute value, $\mathbf{\Pi}_{\xi} = \boldsymbol{\xi}$, and $\mathcal{J} = \mathcal{J}(\mathbf{\Pi}, \vec{\boldsymbol{\lambda}}, \boldsymbol{\xi})$ is the determinant of the Jacobian matrix of the transformation $\mathbf{\Pi} = g(\vec{\boldsymbol{\lambda}}, \boldsymbol{\xi})$. Similarly to Step 1, $\vec{\boldsymbol{\lambda}}^{(t)}$, $\boldsymbol{\xi}^{(t)}$ and $\lambda^{(t)}$ are used to denote the values of the corresponding parameters in the current iteration t of the algorithm.
6. If the move is accepted, then set $\mathbf{\Pi}^{(t+1)} = \mathbf{\Pi}'$, $\boldsymbol{\xi}^{(t+1)} = \boldsymbol{\xi}'$, and $\vec{\boldsymbol{\lambda}}^{(t+1)} = \vec{\boldsymbol{\lambda}}'$ otherwise set $\mathbf{\Pi}^{(t+1)} = \mathbf{\Pi}^{(t)}$ and $\vec{\boldsymbol{\lambda}}^{(t+1)} = \vec{\boldsymbol{\lambda}}^{(t)}$.

The pseudo-parameter vector $\boldsymbol{\xi}$ is used to maintain dimension balance between parameterizations. Therefore, without loss of generality, we can eliminate their effect by assuming that they are uniformly distributed on the zero-one interval, and the ratio $f(\boldsymbol{\xi}')/f(\boldsymbol{\xi}^{(t)})$ disappears from the previous equation.

A Probability-Based Independence Sampler (PBIS) approach is employed for efficient posterior sampling. As detailed in Ntzoufras et al. (2019), the previous general algorithm can be simplified relying on the conditional approach of Ntzoufras and Tarantola (2013). Using $q(\mathbf{\Pi}'|\mathbf{\Pi}^{(t)}) = f_q(\mathbf{\Pi}'|\boldsymbol{n}^{\mathcal{A}})f(\boldsymbol{n}^{\mathcal{A}}|\mathbf{\Pi}^{(t)}, \boldsymbol{n})$, as the proposal distribution, and using as "prior" $f_q(\mathbf{\Pi})$ a product of Dirichlet distributions we obtain a conjugate posterior $f_q(\mathbf{\Pi}|\boldsymbol{n}^{\mathcal{A}})$.

The acceptance rate simplifies to

$$
A = \frac{f\left(\boldsymbol{n}^{\mathcal{A}(t)}|\mathbf{\Pi}'\right)f(\vec{\boldsymbol{\lambda}})f_q\left(\mathbf{\Pi}^{(t)}|\boldsymbol{n}^{\mathcal{A}(t)}\right)}{f\left(\boldsymbol{n}'^{\mathcal{A}}|\mathbf{\Pi}^{(t)}\right)f(\vec{\boldsymbol{\lambda}}^{(t)})f_q\left(\mathbf{\Pi}'|\boldsymbol{n}'^{\mathcal{A}}\right)} \times \mathrm{abs}\left(\frac{\mathcal{J}(\mathbf{\Pi}^{(t)}, \vec{\boldsymbol{\lambda}}^{(t)}, \boldsymbol{\xi}^{(t)})}{\mathcal{J}(\mathbf{\Pi}', \vec{\boldsymbol{\lambda}}', \boldsymbol{\xi}')}\right) \quad (4.17)
$$

We can further simplify PBIS by using the following two-step procedure:

Step 1: Run the Gibbs sampler of Ntzoufras and Tarantola (2013) to obtain a sample from $\mathbf{\Pi}$.

Step 2: Use the sample of step 1 (or sub-sample of it) as a proposal in the general MH algorithm.

Since the sample from Step 1 is generated using an MCMC algorithm, it naturally exhibits serial autocorrelation. To address this issue, we propose to randomly permute the entire MCMC sample before using it as a proposal in the second MCMC step. While this strategy does not completely eliminate autocorrelation, it effectively disrupts the dependence structure between consecutive elements in the proposal sample. More importantly, because the reordered sample is only used as a proposal, the final accepted sample is determined by the MH acceptance rule, which selectively accepts or rejects proposed values. As a result, the final sequence of accepted values does not inherit the correlation structure of the original MCMC sample, further reducing the impact of initial autocorrelation. Although this approach intuitively provides an approximation of the target posterior distribution, we are unaware of any formal mathematical proof to substantiate this claim.

Using such sample (or sub-sample) as proposed values within the MH algorithm is equivalent to using the posterior distribution $f_q(\Pi|n)$ as proposal in the general algorithm, that is $q(\Pi'|\Pi^{(t)}) = f_q(\Pi'|n)$. Under this proposal, the acceptance ratio simplifies to

$$A = \frac{f\left(\vec{\lambda}\right) f_q(\Pi^{(t)})}{f\left(\vec{\lambda}^{(t)}\right) f_q(\Pi')} \times \text{abs}\left(\frac{\mathcal{J}\left(\Pi^{(t)}, \vec{\lambda}^{(t)}, \xi^{(t)}\right)}{\mathcal{J}\left(\Pi', \vec{\lambda}, \xi'\right)}\right).$$

This algorithm is named Prior Adjusted (PA) algorithm.

Example 4.8 (Torus Mandibularis' data) We illustrate our methodology using a dichotomized version of data from Muller and Mayhall (1971), examining the prevalence of torus mandibularis among Inuit groups representing Alaska native groups. This bony protrusion on the mandible is often studied by anthropologists for population and intra-population differences. Previous analyses by Bishop et al. (1975) and Lupparelli (2006) utilized this dataset. Table 4.4 collects data on Age A, Torus mandibularis incidence I, Sex S, and Population P. Inuk groups studied represent distinct regions: Igloolik and Hall Beach from Foxe Basin, and Aleut from Western

Table 4.4 Torus Mandibularis in Inuik populations

Population (P)	Sex (S)	Incidence (I)	Age groups (A)	
			1–20	Over 20
Igloolik and Hall Beach	Male	Present	19	73
		Absent	103	38
	Female	Present	16	61
		Absent	87	36
Aleut	Male	Present	6	18
		Absent	19	14
	Female	Present	4	10
		Absent	17	20

Fig. 4.13 Left panel: bi-directed graph for Torus Mandibularis data; right panel: a Markov equivalent regression graph

Alaska. The Aleut data, collected separately, prompted us to combine Igloolik and Hall Beach groups. Additionally, Age has been median dicothomized.

We examine the four-chain graph depicted in Fig. 4.13, left panel, since it is the one that fits well the original data. The right panel of the same figure includes a Markov equivalent to a regression graph with a different interpretation aimed to

Table 4.5 Posterior summaries and MLEs for marginal log-linear interactions for the Torus Mandibularis data

	PA		RW-λ		ML	
	Mean	SD	Mean	SD	Estimate	SE
$\lambda_{\emptyset}^{AP}$	-1.391	0.004	-1.391	0.004		
λ_{A}^{AP}	-0.001	0.042	-0.003	0.043	-0.002	0.043
λ_{P}^{AP}	-0.072	0.043	-0.079	0.043	-0.072	0.043
λ_{AP}^{AP}	0.000	0.000	0.000	0.000	0.000	
λ_{S}^{AS}	-0.697	0.053	-0.695	0.055	-0.699	0.054
λ_{AS}^{AS}	0.000	0.000	0.000	0.000	0.000	
λ_{I}^{IS}	0.234	0.045	0.241	0.044	0.232	0.044
λ_{IS}^{IS}	0.000	0.000	0.000	0.000	0.000	
λ_{PS}^{APS}	0.004	0.053	-0.009	0.055	0.003	0.054
λ_{APS}^{APS}	0.000	0.000	0.000	0.000	0.000	
λ_{AI}^{AIS}	-0.509	0.051	-0.505	0.052	-0.507	0.051
λ_{AIS}^{AIS}	0.000	0.000	0.000	0.000	0.000	
λ_{IP}^{AIPS}	0.057	0.058	0.082	0.063	0.052	0.062
λ_{AIP}^{AIPS}	0.132	0.068	0.049	0.065	0.151	0.062
λ_{ISP}^{AIPS}	0.029	0.041	0.066	0.063	0.072	0.062
λ_{AIPS}^{AIPS}	0.047	0.046	0.034	0.063	0.037	0.062

study the effect of sex and age on the joint response given by the population and the incidence of torus mandibularis.

Table 4.5 presents the posterior means and standard deviations for the marginal log-linear interactions, achieved through 10,000 iterations and a burn-in period of 1,000, using the proposed PA algorithm and a simple Random Walk (RW-λ) algorithm. Additionally, for comparative analysis, MLEs and their approximate standard errors are included. Analysis of Table 4.5 reveals that the posterior estimates derived from both MCMC methods and the MLEs align for all interactions and main effects in scenarios not involving latent variables, that is for $\lambda_{\text{IP}}^{\text{AIPS}}$, $\lambda_{\text{IPS}}^{\text{AIPS}}$, and $\lambda_{\text{AIPS}}^{\text{AIPS}}$. Here, we note a proportional reduction in standard deviations by 6.5, 34, and 26%, respectively. This outcome aligns with expectations, as the PA algorithm accurately navigates the correct posterior distribution that adheres to the space of parameterizations yielding compatible marginal probabilities. On the other hand, the interaction $\lambda_{\text{AIP}}^{\text{AIPS}}$ demonstrates notable differences, where the RW-λ method's posterior means significantly deviate from the MLEs, whereas the PA method's results are much closer, albeit with a slightly higher standard deviation compared to both the RW-λ outcomes and the MLE standard errors. $\square$

4.6 Concluding Remark

The methodologies presented in this chapter, though primarily focused on bi-directed graphical models, are supported by a flexible framework that allows for their extension to regression graph models Markov equivalent to either DAGs or bi-directed graphs. This adaptability facilitates their application across a diverse array of graphical configurations, significantly enhancing the utility of the Bayesian approaches discussed.

In scenarios where the regression graph is Markov equivalent to a DAG, conjugate analysis becomes not only feasible but also highly advantageous. This analysis leverages the structural properties of the DAG to facilitate simpler and more direct computations. The use of conjugate priors within this framework simplifies computational processes and boosts analytical efficiency, making it an especially effective approach for these graph types.

For regression graphs that are Markov equivalent to a bi-directed graph and do not admit a DAG representation with only observed variables, a representation involving an augmented DAG with latent variables should be employed. This adaptation is essential for addressing more complex graphical models where direct conjugate analysis is unfeasible. By integrating latent variables into the Bayesian framework, it becomes possible to effectively manage the added complexities introduced by the absence of DAG equivalence, thereby providing an analytical tool for these intricate scenarios.

4.7 Appendix: Jacobian and Link Function

4.7.1 Jacobian Matrix

Let us consider the mapping $\boldsymbol{\Pi} \mapsto \boldsymbol{\lambda}^{\mathcal{A}} = (\overrightarrow{\boldsymbol{\lambda}}, \boldsymbol{\xi})$. The terms of the Jacobian matrix of this mapping are $\mathcal{J}(\boldsymbol{\Pi}, \overrightarrow{\boldsymbol{\lambda}}, \boldsymbol{\xi}) = \left| \dfrac{\partial \boldsymbol{\Pi}}{\partial (\overrightarrow{\boldsymbol{\lambda}}, \boldsymbol{\xi})} \right|$. Each terms can be expressed in a simplified version as

$$
\mathcal{J}^{-1} = \left| \frac{\partial (\overrightarrow{\boldsymbol{\lambda}}, \boldsymbol{\xi})}{\partial \boldsymbol{\Pi}} \right| = \left| \frac{\partial (\overrightarrow{\boldsymbol{\lambda}}, \boldsymbol{\xi})}{\partial (\boldsymbol{\Pi}_\xi, \boldsymbol{\Pi}_{\backslash\xi})} \right| = \left| \begin{matrix} \dfrac{\partial \overrightarrow{\boldsymbol{\lambda}}}{\partial \boldsymbol{\Pi}_\xi} & \dfrac{\partial \overrightarrow{\boldsymbol{\lambda}}}{\partial \boldsymbol{\Pi}_{\backslash\xi}} \\[2mm] \dfrac{\partial \boldsymbol{\xi}}{\partial \boldsymbol{\Pi}_\xi} & \dfrac{\partial \boldsymbol{\xi}}{\partial \boldsymbol{\Pi}_{\backslash\xi}} \end{matrix} \right|
$$

$$
= \left| \begin{matrix} \dfrac{\partial \overrightarrow{\boldsymbol{\lambda}}}{\partial \boldsymbol{\Pi}_\xi} & \dfrac{\partial \overrightarrow{\boldsymbol{\lambda}}}{\partial \boldsymbol{\Pi}_{\backslash\xi}} \\[2mm] \boldsymbol{I} & \boldsymbol{0} \end{matrix} \right| = - \left| \frac{\partial \overrightarrow{\boldsymbol{\lambda}}}{\partial \boldsymbol{\Pi}_{\backslash\xi}} \right| ,
$$

where $\boldsymbol{\Pi}_{\backslash\xi}$ is obtained from $\boldsymbol{\Pi}$ after excluding the elements of $\boldsymbol{\Pi}_\xi$. Then, the elements of matrix $\mathcal{J}(\boldsymbol{\Pi}, \overrightarrow{\boldsymbol{\lambda}}, \boldsymbol{\xi})$ are given by

$$
\frac{\partial \lambda_k}{\partial \Pi_j} = \sum_{l=1}^{\mathrm{col}(\boldsymbol{C})} \left\{ C_{kl} \left(\sum_{\iota=1}^{|\mathcal{I}|} M_{l\iota} \pi_\iota \right)^{-1} \sum_{\iota=1}^{|\mathcal{I}|} M_{l\iota} \Delta_{\iota j} \right\} \quad \text{with } \Delta_{\iota j} = \frac{\partial \pi_\iota}{\partial \Pi_j}, \qquad (4.18)
$$

where π_ι denote the ι-element of $\boldsymbol{\pi}$, and $\mathrm{col}(\boldsymbol{C})$ is the number of columns of the contrast matrix $\boldsymbol{C}$. Equation (4.18) can be rewritten as

$$
\frac{\partial \lambda_k}{\partial \Pi_j} = \sum_{l=1}^{\mathrm{col}(\boldsymbol{C})} Q_{klj} \quad \text{with } Q_{klj} = C_{kl} \frac{\sum_{\iota=1}^{|\mathcal{I}|} M_{l\iota} \Delta_{\iota j}}{\sum_{\iota=1}^{|\mathcal{I}|} M_{l\iota} \pi_\iota} ,
$$

where $\boldsymbol{M} = (M_{l\iota})$ is a $\mathrm{col}(\boldsymbol{C}) \times |\mathcal{I}|$ matrix, $\boldsymbol{\Delta} = (\Delta_{\iota j})$ is a matrix of dimension $|\mathcal{I}| \times \dim(\boldsymbol{\Pi})$, and $\dim(\boldsymbol{\Pi})$ is the dimension of the probability vector $\boldsymbol{\Pi}$.

We now describe how to obtain the derivative terms $\Delta_{\iota j}$. Let us now denote by i' the index of vector $\boldsymbol{\pi}$ such that $\pi_{i'} \equiv p(i)$. Moreover, the index j corresponds to a variable $u_j \in \mathcal{A}$ such that $\Pi_j \equiv \pi_{u_j | \mathrm{pa}(u_j)}(j_{u_j} | j_{\mathrm{pa}(u_j)})$ for a specific cell j of the augmented table $\mathcal{I}^{\mathcal{A}}$. Therefore, for any $\iota = i'$, terms $\Delta_{\iota j}$ are given by

$$
\Delta_{\iota j} = \Delta_{i' j} = \frac{\partial \pi_i'}{\partial \Pi_j} = \frac{\partial p(i)}{\partial \pi_{u_j | \mathrm{pa}(u_j)}(j_{u_j} | j_{\mathrm{pa}(u_j)})} \quad \text{for every } i' \mapsto i \text{ and } j \mapsto (u_j, j).
$$

$$
(4.19)
$$

For the computation of each $\Delta_{\iota j}$ we consider two different cases: (A) $u_j \in \mathcal{V}$ and (B) $u_j \in \mathcal{L}$. In the following, to simplify notation, we denote u_j by u. Furthermore, we indicate with $\mathcal{L}_u = \mathcal{L} \cap \mathrm{pa}(u)$ the latent variables that are parents of u, and with $\mathcal{A}_u = \mathcal{V} \cup \{u\} \cup \mathcal{L}_u$.

For **case A**, when u is an observed variable, we obtain

$$\frac{\partial p(i)}{\partial \pi_{u|\mathrm{pa}(u)}\left(j_u \mid j_{\mathrm{pa}(u)}\right)} = \delta(i, j) \frac{p^{\mathcal{A}_u}\left(i, j_{\mathcal{L}_u}\right)}{\pi_{u|\mathrm{pa}(u)}\left(i_u \mid j_{\mathrm{pa}(u)}\right)}, \tag{4.20}$$

with

$$\delta(i, j) = \begin{cases} 1 & \text{if } i_u = j_u < |\mathcal{I}_u| \text{ and } i_{pa(u)\backslash\mathcal{L}} = j_{pa(u)\backslash\mathcal{L}} \\ -1 & \text{if } j_u \neq i_u = |\mathcal{I}_u| \text{ and } i_{pa(u)\backslash\mathcal{L}} = j_{pa(u)\backslash\mathcal{L}} \\ 0 & \text{if } j_u \neq i_u < |\mathcal{I}_u| \text{ or } i_{pa(u)\backslash L} \neq j_{pa(u)\backslash\mathcal{L}} \end{cases}, \tag{4.21}$$

where

$$p^{\mathcal{A}_u}\left(i, j_{\mathcal{L}_u}\right) = \begin{cases} P\left(Y_{\mathcal{V}} = i, Y_{\mathcal{L}\cap\mathrm{pa}(u)} = j_{\mathcal{L}\cap\mathrm{pa}(u)}\right) & \mathcal{L}_u \neq \emptyset \\ p(i) & \mathcal{L}_u = \emptyset \end{cases}.$$

For **case B**, when u is a latent variable, $\mathrm{pa}(u) = \emptyset$ due to the structure of the DAG representation. Hence, the derivative is given by

$$\frac{\partial p(i)}{\partial \pi_{u|pa(u)}\left(j_u \mid j_{pa(u)}\right)} = \frac{\partial p(i)}{\partial \pi_u\left(j_u\right)} = \frac{p^{\mathcal{A}_u}(i, j_u)}{\pi_u(j_u)} - \frac{p^{\mathcal{A}_u}\left(i, |\mathcal{I}_u|\right)}{\pi_u\left(|\mathcal{I}_u|\right)} \text{ for } j_u < |\mathcal{I}_u|. \tag{4.22}$$

Once $\boldsymbol{\Delta}$ is obtained we can construct the Jacobian matrix using the following steps

1. construct $\boldsymbol{H} = \boldsymbol{M}\boldsymbol{\Delta}$ a matrix of dimension $\mathrm{col}(\boldsymbol{C}) \times \dim(\boldsymbol{\Pi})$;
2. construct $\boldsymbol{\Gamma} = \boldsymbol{M}\boldsymbol{P}$ vector of dimension $\mathrm{col}(\boldsymbol{C}) \times 1$;
3. construct $\boldsymbol{\Gamma}' = \boldsymbol{\Gamma}\mathbf{1}_{\dim(\boldsymbol{\Pi})}^{\mathsf{T}}$, with $\mathbf{1}_{\dim(\boldsymbol{\Pi})}$ a $\dim(\boldsymbol{\Pi}) \times 1$ vector of ones, and $\boldsymbol{\Gamma}'$ a matrix of dimension $\mathrm{col}(\boldsymbol{C}) \times \dim(\boldsymbol{\Pi})$ with all columns equal to $\boldsymbol{\Gamma}$;
4. set $\boldsymbol{H}' = \boldsymbol{H} \circ \boldsymbol{\Gamma}''$ where $\circ$ indicates the Hadamard product (element by element multiplication), and $\boldsymbol{\Gamma}''$ is a matrix with elements $\Gamma''_{\nu\kappa} = 1/\Gamma'_{\nu\kappa}$;
5. denote with $\boldsymbol{J} = \boldsymbol{C}\boldsymbol{H}'$ the Jacobian matrix dimension $\dim(\boldsymbol{\Pi}) \times \dim(\boldsymbol{\Pi})$.

For more details see Sect. 4.4 of Ntzoufras et al. (2019).

4.7.2 Contrast Matrix $C_{\mathcal{M}}$ with Sum to Zero Constraints

Firstly we need to construct the design matrix $\mathbf{X}_{\mathcal{V}}$ for the saturated model of the cross-classification of discrete variables $Y_{\mathcal{V}}$ with sum to zero constraints. For each variable v we have the following matrix

$$\mathbf{J}_v = \begin{bmatrix} 1 & -\mathbf{1}_{(|\mathcal{I}_v|-1)}^{\mathsf{T}} \\ \mathbf{1}_{(|\mathcal{I}_v|-1)} & \mathbf{I}_{(|\mathcal{I}_v|-1)} \end{bmatrix},$$

where $\mathbf{1}_\kappa$ is a vector of ones of length κ while $\mathbf{I}_\kappa$ is the identity matrix of dimension $\kappa \times \kappa$.

The design matrix of the saturated model will be of dimension $\left(\prod_{v \in \mathcal{V}} |\mathcal{I}_v|\right) \times \left(\prod_{v \in \mathcal{V}} |\mathcal{I}_v|\right)$ and can be obtained as

$$\mathbf{Y}_\mathcal{V} = \bigotimes_{q=0}^{|\mathcal{V}|-1} \mathbf{J}_{|\mathcal{V}|-q}.$$

The contrast matrix $\mathbf{C}$ can be constructed by using the following rules.

1. For each margin M_m, we construct the design matrix $\mathbf{X}_{M_m}$ corresponding to the saturated model (using sum-to-zero constraints) of the cross-classification of variables Y_{M_m}. Then we consider its inverse $\mathbf{X}_{M_m}^{-1}$ in order to obtain the corresponding contrast matrix. Now, let $\mathbf{C}_m$ be a sub-matrix of the contrast matrix $\mathbf{X}_{M_m}^{-1}$ obtained by deleting rows corresponding to interactions that are not obtained from margin M_m.
2. The contrast matrix $\mathbf{C}$ is obtained as

$$\mathbf{C} = \bigoplus_{m:\ M_m \in \mathcal{M}} \mathbf{C}_m = \mathrm{diag}(C_1, \ldots, C_{|\mathcal{M}|}),$$

where $\bigoplus$ denotes the matrix direct sum.

4.8 Bibliographic Notes

Section 4.1

This section introduces Bayesian analysis for parametric inference and model comparison. While not exhaustive, the key references on the topic include Robert (2007), Bernardo and Smith (2009) and Gelman et al. (2013). Specifically, the learning mechanism based on combination of information provided by prior knowledge and observed data is illustrated and, within the context of model comparison, the relevance of prior compatibility is discussed. For additional insights into compatible prior distributions, refer to Dawid and Lauritzen (2001), Roverato and Consonni (2004), and Consonni and Veronese (2008).

Section 4.2

The Bayesian analysis of regression graph models for categorical data has not been developed as much as Frequentist methods. Some context specific results have been presented by e.g., Silva and Ghahramani (2009b), Bartolucci et al. (2012) and Ntzoufras and Tarantola (2013) who propose an approach for Bayesian inference exploiting the Markov equivalence between regression graphs, bi-directed graphs and the augmented DAG representation. This method is based on the use of product of Dirichlet distribution as a prior distribution; see Heckerman et al. (1995). A procedure for quantitative learning of marginal log-linear parameters has been proposed by Ntzoufras et al. (2019).

Section 4.3

The Markov equivalence between regression graphs, DAGs and bi-directed graphs can be automatically verified by using the function `MarkEqRcg` in R package **ggm** (Marchetti et al. 2024). See also the short tutorial in Sadeghi and Marchetti (2012) illustrating the functions in the package that deal with mixed graphs. These graphs are important because they capture the modified independence structure after marginalization over, and conditioning on, nodes of directed acyclic graphs. The terminology homogeneous and non-homogeneous has been introduced by Ntzoufras and Tarantola (2013) adapting to bi-directed graphs the notation of Letac and Massam (2007).

For conditions under which homogeneous and non-homogeneous models are compatible with an augmented DAG representation see Pearl and Wermuth (1994), Drton and Richardson (2008), Silva and Ghahramani (2009a, 2009b). Although every bi-directed graph can be graphically represented using an augmented DAG, the model derived through marginalization may not be directly equivalent to the original one. Specifically, the introduction of latent variables can result in extra inequality constraints on the observed variables' distribution. An illustrative example is the 4-chain bi-directed graph, which, when augmented with the inclusion of an additional latent variable, requires the marginal distribution of the observables to conform to Bell's inequalities; see, for example, Tian and Pearl (2002), Evans (2018), Kang and Tian (2006). Therefore, the augmented DAG can be considered as a close approximation of the corresponding marginal association model unless all inequalities become inactive a-posteriori (i.e., are satisfied by the posterior distribution of the probabilities of the marginal association model).

Section 4.4

The selection of hyperparameters values for the prior distribution is critically important for model comparison, given the well-documented sensitivity of posterior model odds and the impact of the Bartlett-Lindley paradox. This paradox arises when the Bayesian evidence disproportionately favors a model as priors become increasingly non-informative, effectively overriding the actual data observed. In the context of model comparisons, this phenomenon often leads to a paradoxical preference for simpler models, regardless of the data presented. This underscores the necessity of carefully choosing prior parameters to ensure that the model comparison genuinely reflects the evidence provided by the data. For a detailed discussion see Sect. 11.2 in Ntzoufras (2008). A detailed comparison of prior choices is presented in Table 2 of Ntzoufras and Tarantola (2013).

The median probability model, is the one consisting of those edges that have overall posterior probability greater than or equal to 0.5 of being present in the model. Under the Bayesian approach, it is commonly perceived that the optimal predictive model is the model with highest posterior probability, but this is not necessarily the case. Starting from this consideration Barbieri and Berger (2004) show that, for selection among normal linear models, the optimal predictive model is often the median probability model that in some case does not correspond with the highest probability model.

A detailed description on how to implements Chib's marginal likelihood estimator for non-homogeneous models is provided in Sect. 4.4 of Ntzoufras and Tarantola (2013). Due to the label switching problem, Chib's marginal likelihood estimator is adjusted applying the correction originally proposed Neal (1998) and further developed in more details by Marin and Robert (2008). Moreover, the mode (or values close to it) is most suitable choice for $\pi^{*\mathcal{D}}$ that can be used in Chib's estimator since the mean and the median will be away from points of high posterior density if the MCMC explores all local modes. In cases that the MCMC visits only one of the permutations of the labels of the latent variables, then using the posterior mean and median in Chib's estimator also results in good estimates of the marginal likelihood.

Finally a detailed presentation on Bayed Factor construction and its properties is provided in Kass and Raftery (1995).

Section 4.5

For more details regarding variation independence in marginal log-linear models see Lupparelli et al. (2009) and Evans and Richardson (2013). A detailed discussion of the MCMC strategy discussed here is presented in Ntzoufras et al. (2019).

References

Aitchison, J., & Silvey, S. D. (1958). Maximum likelihood estimation of parameter subject to restraints. *Annals of Mathematical Statistics, 29*, 691–711.

Akaike, H. (1973). Information theory and an extension of the maximum likelihood principle. In B. N. Petrov & F. B. Csaki (Eds.), *Second international symposium on information theory* (pp. 267–281). Budapest: Akademiai Kiado.

Anderson, T. W. (1973). Asymptotically efficient estimation of covariance matrices with linear structure. *Annals of Statistics, 1*, 135–141.

Barbieri, M. M., & Berger, J. O. (2004). Optimal predictive model selection. *Annals of Statistics, 32*(3), 870–897.

Barndorff-Nielsen, O. (1978). *Information and exponential families in statistical theory*. New York: Wiley.

Bartolucci, F., Colombi, R., & Forcina, A. (2007). An extended class of marginal link functions for modelling contingency tables by equality and inequality constraints. *Statistica Sinica, 17*, 813–828.

Bartolucci, F., Scaccia, L., & Farcomeni, A. (2012). Bayesian inference through encompassing priors and importance sampling for a class of marginal models for categorical data. *Computational Statistics & Data Analysis, 56*, 4067–4080.

Bergsma, W. P., & Rudas, T. (2002). Marginal log-linear models for categorical data. *Annals of Statistics, 30*(1), 140–159.

Bergsma, W., Croon, M., & Hagenaars, J. A. (2009). *Marginal models for dependent, clustered, and longitudinal categorical data*. London, UK: Springer.

Bergsma, P. W. (1997). *Marginal models for categorical variables*. Tilburg: University Press.

Bernardo, J. M., & Smith, A. F. M. (2009). *Bayesian theory*. Wiley Series in Probability and Statistics. Wiley.

Bishop, Y. M., Fienberg, S. E., & Holland, P. W. (1975). *Discrete multivariate analysis*. Cambridge: MIT Press.

Chib, S. (1995). Marginal likelihood from the Gibbs output. *Journal of the American Statistical Association, 90*(432), 1313–1321.

Colombi, R., & Forcina, A. (2001). Marginal regression models for the analysis of positive association of ordinal response variabels. *Biometrika, 88*(4), 1007–1019.

Colombi, R., Giordano, S., & Cazzaro, M. (2014). HMMM: An R package for hierarchical multinomial marginal models. *Journal of Statistical Software, 59*, 1–25.

Consonni, G., & Veronese, P. (2008). Compatibility of prior specifications across linear models. *Statistical Science, 23*(3), 332–353.

© The Editor(s) (if applicable) and The Author(s), under exclusive license to Springer 103
Nature Switzerland AG 2025

M. Lupparelli et al., *Regression Graph Models for Categorical Data*,
SpringerBriefs in Statistics, https://doi.org/10.1007/978-3-031-99797-6

Coppen, A. (1966). The Marke-Nyman temperament scale: an English translation. *British Journal of Medical Psychology, 39*(1), 55–59.

Cox, D. R., & Wermuth, N. (1993). Linear dependencies represented by chain graphs (with discussion). *Statistical Science, 8*(204–218), 247–277.

Cox, D. R., & Snell, E. J. (1989). *Analysis of binary data* (2nd ed.). London: Chapman and Hall.

Cox, D., R., & Wermuth, N. (1996). *Multivariate dependencies. Models, analysis and interpretation.* London: Chapman and Hall.

Davis, J. A., Smith, T. W., & Mardsen, J. A. (2007). *General Social Survey Cumulative Codebook: 1972–2006.* Chicago: NORC.

Dawid, P., & Lauritzen, S. (2001). Compatible prior distributions. In *Bayesian methods* (pp. 109–116). Eurostat.

Dellaportas, P., & Forster, J. J. (1999). Markov chain Monte Carlo model determination for hierarchical and graphical log-linear models. *Biometrika, 86,* 615–633.

Dethlefsen, C., & Højsgaard, S. (2005). A common platform for graphical models in R: The gRbase package. *Journal of Statistical Software, 14*(17), 1–12.

Drton, M. (2009). Discrete chain graph models. *Bernoulli, 15,* 736–753.

Drton, M., & Richardson, T. S. (2008). Binary models for marginal independence. *Journal of the Royal Statistical Society, Series B, 70,* 287–309.

Ekholm, A., Smith, P. W. F., & McDonald, J. W. (1995). Marginal regression analysis of a multivariate binary response. *Biometrika, 82*(4), 847–854.

Evans, R. (2018). Margins of discrete Bayesian networks. *Annals of Statistics, 46,* 2623–2656.

Evans, R. J., & Forcina, A. (2013). Two algorithms for fitting constrained marginal models. *Computational Statistics & Data Analysis, 66,* 1–7.

Evans, R., & Richardson, T. (2013). Marginal log-linear parameters for graphical Markov models. *Journal of the Royal Statistical Society, Series B, 75,* 743–768.

Forcina, A., Lupparelli, M., & Marchetti, G. M. (2010). Marginal parameterizations of discrete models defined by a set of conditional independencies. *Journal of Multivariate Analysis, 101*(10), 2519–2527.

Fox, J., & Weisberg, S. (2018). Visualizing fit and lack of fit in complex regression models with predictor effect plots and partial residuals. *Journal of Statistical Software, 87*(9), 1–27.

Fox, J., & Weisberg, S. (2019). *An R companion to applied regression* (3rd ed.). Los Angeles: Sage Publications.

Gelman, A., Carlin, J. B., Stern, H. S., Dunson, D. B., Vehtari, A., & Rubin, D. B. (2013). *Bayesian data analysis. Chapman & Hall/CRC Texts in Statistical Science* (3rd ed.). New York: Taylor & Francis.

Glonek, G. J. N., & McCullagh, P. (1995). Multivariate logistic models. *Journal of the Royal Statistical Society, Series B, 57*(3), 533–546.

Heckerman, D., Geiger, D., & Chickering, D. (1995). Learning Bayesian networks: The combination of knowledge and statistical data. *Machine Learning, 20,* 197–243.

Javidian, M. A., & Valtorta, M. (2021). A decomposition-based algorithm for learning the structure of multivariate regression chain graphs. *International Journal of Approximate Reasoning, 136,* 66–85.

Kang, C., & Tian, J. (2006). Inequality constraints in causal models with hidden variables. In *Proceedings of the Twenty-Second Conference on Uncertainty in Artificial Intelligence* (pp. 233–240). Cambridge, MA, USA: AUAI Press.

Kass, R. E., & Raftery, A. E. (1995). Bayes factors. *Journal of the American Statistical Association, 90*(430), 773–795.

Kauermann, G. (1997). Note on the multivariate logistic models for contingency table. *Australian Journal of Statistics, 39,* 261–276.

Knuiman, M. W., & Speed, T. P. (1988). Incorporating prior information into the analysis of contingency tables. *Biometrics, 44,* 1061–1071.

La Rocca, L., & Roverato, A. (2018). Discrete graphical models and their parameterizations. In M. Maathius, M. Drton, S. Lauritzen, & M. Wainwright (Eds.), *Handbook of graphical models* (pp. 193–218). Chapman & Hall/CRC.

Lang, J. B. (1996). Maximum likelihood methods for a generalized class of log-linear models. *Annals of Statistics, 24*(2), 726–752.

Lang, J. B. (2005). Homogeneous linear predictor models for contingency tables. *Annals of Statistics, 100*, 121–134.

Lang, J. B., & Agresti, A. (1994). Simultaneously modeling joint and marginal distributions of multivariate categorical responces. *Journal of the American Statistical Association, 89*(426), 625–632.

Lauritzen, S. L. (1996). *Graphical models*. Oxford: Oxford University Press.

Letac, G., & Massam, H. (2007). Wishart distributions for decomposable graphs. *Annals of Statistics, 35*, 1278–1323.

Lienert, G. A. (1970). Konfigurationsfrequenzalyse einiger Lysergsäurediäthylamid-wirkungen. *Arzneimittellorschung, 20*(7), 912–913.

Lupparelli, M. (2006). *Graphical models of marginal independence for categorical data*. Ph.D. thesis, Department of Statistics, University of Florence.

Lupparelli, M., Marchetti, G. M., & Bergsma, W. P. (2009). Parameterizations and fitting of bi-directed graph models to categorical data. *Scandinavian Journal of Statistics, 36*(3), 559–576.

Lupparelli, M., & Roverato, A. (2017). Log-mean linear regression models for binary responses with an application to multimorbidity. *Journal of the Royal Society, Series C, 66*, 227–252.

Marchetti, G. M., Drton, M., & K. Sadeghi (2024). *GGM: Graphical Markov models with mixed graphs. CRAN*. https://CRAN.R-project.org/package=ggm.

Marchetti, G. M., & Lupparelli, M. (2011). Chain graph models of multivariate regression type for categorical data. *Bernoulli, 17*(3), 827–844.

Marin, J. M., & Robert, C. (2008). Approximating the marginal likelihood in mixture models. 1–7. https://doi.org/10.48550/arXiv.0804.2414

Kalisch, M., Mächler, M., Colombo, D., Maathuis, M. H., & Bühlmann, P. (2012). Causal inference using graphical models with the R package pcalg. *Journal of Statistical Software, 47*(11), 1–26.

McCullagh, P., & Nelder, J. A. (1989). *Generalized linear models*. London: Chapman and Hall.

Muller, T. P., & Mayhall, J. T. (1971). Analysis of contingency table data on torus mandibularis using a log linear model. *American Journal of Physical Anthropology, 34*(1), 149–153.

Neal, R. M. (1998). Erroneous results in marginal likelihood from the Gibbs output (with associated comments written 16 March 1999). Technical report, Department of Statistics and Department of Computer Science, University of Toronto.

Ntzoufras, I. (2008). *Bayesian modeling using WinBUGS*. Wiley.

Ntzoufras, I., & Tarantola, C. (2013). Conjugate and conditional conjugate Bayesian analysis of discrete graphical models of marginal independence. *Computational Statistics & Data Analysis, 66*, 161–177.

Ntzoufras, I., Tarantola, C., & Lupparelli, M. (2019). Probability based independence sampler for Bayesian quantitative learning in graphical log-linear marginal models. *Bayesian Analysis, 14*, 777–803.

Pearl, J., & Wermuth, N. (1994). When can association graphs admit a causal interpretation? In P. Cheeseman, & R. W. Oldford (Eds.), *Selecting Models from Data. Lecture Notes in Statistics* (Vol. 89, pp. 205–214). New York, NY: Springer.

Retherford, R. D., & Choe, M. K. (1993). *Statistical models for causal analysis*. New York: Wiley.

Richardson, T. S. (2003). Markov property for acyclic directed mixed graphs. *Scandinavian Journal of Statistics, 30*(1), 145–157.

Richardson, T. S., & Spirtes, P. (2002). Ancestral graph Markov models. *Annals of Statistics, 30*, 962–103.

Robert, C. (2007). *The Bayesian choice: From decision-theoretic foundations to computational implementation. Springer Texts in Statistics*. New York: Springer.

Roverato, A. (2017). *Graphical models for categorical data.* Cambridge, UK: Cambridge University Press.

Roverato, A., & Consonni, G. (2004). Compatible prior distributions for DAG models. *Journal of the Royal Statistical Society, Series B, 66,* 47–61.

Roverato, A., Lupparelli, M., & La Rocca, L. (2013). Log-mean linear models for binary data. *Biometrika, 100,* 485–494.

Rudas, T. (1998). *Odds ratio in the analysis of contingency tables.* Thousand Oaks: Sage.

Rudas, T., Bergsma, W., & Nemeth, R. (2010). Marginal log-linear parameterization of conditional independence models. *Biometrika, 97*(4), 1006–1012.

Sadeghi, K. (2013). Stable mixed graphs. *Bernoulli, 19*(5B), 2330–2358.

Sadeghi, K., & Marchetti, G. M. (2012). Graphical markov models with mixed graphs in R. *The R Journal, 4*(2), 65–73. https://doi.org/10.32614/RJ-2012-015.

Scutari, M. (2010). Learning Bayesian networks with the bnlearn R package. *Journal of Statistical Software, 35*(3), 1–22.

Scutari, M., & Denis, J.-B. (2021). *Bayesian networks with examples in R* (2nd ed.). Boca Raton: Chapman and Hall. ISBN 978-0367366513.

Silva, R., & Ghahramani, Z. (2009a). Factorial mixture of Gaussians and the marginal independence model. In D. van Dyk & M. Welling (Eds.), *Proceedings of the Twelfth International Conference on Artificial Intelligence and Statistics.* (Vol. 5, pp. 520–527).

Silva, R., & Ghahramani, Z. (2009). The hidden life of latent variables: Bayesian learning with mixed graph models. *Journal of Machine Learning Research, 10,* 1187–1238.

Sonntag, D., & Peña, J. M. (2012). Learning multivariate regression chain graphs under faithfulness. In *Sixth European Workshop on Probabilistic Graphical Models* (pp. 299–306).

Studený, M. (2005). *Probabilistic conditional independence structures.* London, UK: Springer.

Tian, J., & Pearl, J. (2002). On the testable implications of causal models with hidden variables. In A. Darwiche, & N. Friedman (Eds.), *Proceedings of the Eighteenth Conference on Uncertainty in Artificial Intelligence* (pp. 519–527). San Francisco, CA: Morgan Kaufmann.

Wasserman, L. (2006). *All of statistics.* New York: Springer.

Wermuth, N. (2011). Probability distributions with summary graph structure. *Bernoulli, 17*(3), 845–879.

Wermuth, N. (2015). Graphical Markov models, unifying results and their interpretation. In N. Balakrishnan (Ed.), *Wiley StatsRef: Statistics reference online* (pp. 1–34). New York: Wiley.

Wermuth, N., & Cox, D. (1995). Derived variables calculated from similar responses: Some characteristics and examples. *Computational Statistics and Data Analysis, 19,* 223–234.

Wermuth, N., & Sadeghi, K. (2012). Sequences of regressions and their independences. *TEST, 21,* 215–252.

Whittaker, J. (1990). *Graphical models in applied multivariate statistics.* Chichester: Wiley.

Wilkinson, G. N., & Rogers, C. E. (1973). Symbolic description of factorial models for analysis of variance. *Journal of the Royal Statistical Society, Series B, 22,* 392–399.

Zellner, A. (1962). An efficient method of estimating seemingly unrelated regressions and tests for aggregation bias. *Journal of the American Statistical Association, 57*(298), 348–368.

Index

M. Lupparelli et al., *Regression Graph Models for Categorical Data*,

SpringerBriefs in Statistics, https://doi.org/10.1007/978-3-031-99797-6

MIX
Papier aus verantwortungsvollen Quellen
Paper from responsible sources
FSC® C105338

If you have any concerns about our products,
you can contact us on
ProductSafety@springernature.com

In case Publisher is established outside the EU,
the EU authorized representative is:
Springer Nature Customer Service Center GmbH
Europaplatz 3, 69115 Heidelberg, Germany

Printed by Libri Plureos GmbH
in Hamburg, Germany